AF334539

# ABIOTIC STRESS

# ROLE IN SUSTAINABLE AGRICULTURE, DETRIMENTAL EFFECTS AND MANAGEMENT STRATEGIES

# ENVIRONMENTAL HEALTH - PHYSICAL, CHEMICAL AND BIOLOGICAL FACTORS

Additional books in this series can be found on Nova's website under the Series tab.

Additional e-books in this series can be found on Nova's website under the e-book tab.

Environmental Health - Physical, Chemical and Biological Factors

# Abiotic Stress

## Role in Sustainable Agriculture, Detrimental Effects and Management Strategies

Annabella Ferro

Editor

nova publishers

New York

## NOTICE TO THE READER



**Library of Congress Cataloging-in-Publication Data**

ISBN: 978-1-63117-622-7

*Published by Nova Science Publishers, Inc. † New York*

# CONTENTS

# PREFACE

Stress may be defined as any negative effect or condition that a living organism may suffer. Abiotic stresses, such as salinity, drought, high and low temperatures, ozone, heavy metals, low concentration of nutrients in degraded soils, can cause deleterious effects in almost all phonological plant stages, from germination to full plant development. It has been estimated that 70% of the crop yield loss can be attributed to abiotic stresses, especially drought. It is known that plants respond readily when challenged by abiotic stresses by the modified regulation of many genes. This is a manifold response consisting of transcriptional and translational mechanisms that allow the adaptation of plants to many unfavorable environments. This book discusses the regulation of abiotic stress responses in plants; the structural aspects and functional regulation of late embryogenesis abundant genes and proteins conferring abiotic stress tolerance in plants; and dehydration responsive element binding transcription factors.

Chapter 1 – Abiotic stresses have become one the principal factors limiting crop productivity world-wide. Abiotic stresses like salinity, drought, high and low temperatures, ozone, heavy metals, and low concentration of nutrients in degraded soils, cause deleterious effects in almost all phenological plant stages, from germination to full plant development. It is known that plants respond readily when challenged by abiotic stresses through the modified regulation of many genes. This is a manifold response consisting of transcriptional and translational mechanisms that allow the adaptation of plants to many unfavorable environments.

To date, many genetic and molecular strategies have been developed to identify those factors that are important for plants to adapt to stressful environments. As a result, a comprehensive knowledge regarding the pattern of induction of many genes during adverse conditions, have uncovered a complex network of inter-connected responses. At a further level of regulation, the function of transcription factors has been found to be essential to regulate the proper activation of an adequate answer to given environmental stimuli consisting of specific posttranslational and epigenetic modifications, including changes in nucleosome distribution, histone modification, DNA methylation, and npcRNAs (non-protein-coding RNAs) that play an important role in abiotic stress gene networks as well.

In the post-genomics era, data garnered from comprehensive analyses using increasingly efficient technologies such as transcriptomics, proteomics and metabolomics, have increased the authors understanding of the complex regulatory networks associated with stress adaptation and tolerance; besides, natural variations for stress responses have been observed at different levels of integration using strategies like classical linkage and association (LD) mapping.

Moreover, classical breeding approaches have revealed that stress tolerance traits are dispersed in various quantitative trait loci (QTLs), a factor which hinders their rapid genetic selection. Geneticists are working on fundamental mechanisms of adaptation which could be useful to crop breeders looking for natural variants that could optimize environmental resources and provide targets for breeding programs.

Taking together all of the above aspects, this chapter describes some of the most recent investigations about molecular, physiological and biochemical changes that are triggered in response to abiotic stress in plants. Its main message is that in order to deal with changing and challenging environmental conditions, plants exhibit a wide range of integrated responses which usually display complex quantitative variation.

Chapter 2 – Late embryogenesis abundant (LEA) proteins are the members of a large group of hydrophilic, low molecular weight (10–30 kDa) proteins found primarily in plants, encoded by multigene families. They were first characterized in cotton and wheat and are produced in abundance, late during embryo development, constituting around 4% of the total cellular proteins. They are mostly localized in cytoplasm and nuclear region. Genes of LEA proteins have been identified in many plant species; their expression is linked to the acquisition of desiccation tolerance in orthodox seeds, pollen and anhydrobiotic plants. However, many LEA proteins are induced by salinity, drought, cold or other osmotic stress and by exogenous abscisic acid (ABA) in vegetative tissues as well, where they play a major role as cellular protectants. These proteins are part of evolutionarily conserved group of hydrophilic proteins termed "hydrophilins" involved in various adaptive responses to hyperosmotic conditions. Although their precise function is still obscure, a number of putative mechanisms have been proposed to LEA proteins where they act as either hydrating buffer, molecular shield by sequestering ions, as chemical chaperones by helping in renaturing unfolded cellular proteins, and transport of nuclear targeted proteins during stress. The majority of LEA proteins display preponderance of hydrophilic and charged amino acid residues, lacking or having a low proportion of Cys and Trp residues. At least six different groups of LEA proteins have been identified based on their amino acid sequence, mRNA homology and expression pattern; the major categories being group 1, group 2, group 3, group 4 and group 5. LEA protein synthesis, expression and biological activities are regulated by many factors (e.g., developmental stages, hormones, ion change and dehydration) and signal transduction pathways. The best characterized cis-element in LEA genes with context to ABA-mediated osmotic stress induction is the ABA-responsive element (ABRE), which contains the palindromic motif CACGTC. The ABREs can interact mostly with the basic leucine zipper (bZIP) transcription factors. A 9-bp conserved sequence, TACGACAT, termed dehydration responsive element (DRE) or C-repeat (CRT) elements has been reported in the promoter regions of ABA-independent cold- and drought-inducible genes. Transgenic plants overexpressing several LEA genes under constitutive or abiotic stress-inducible promoters have been reported to show enhanced tolerance to single or multiple stress conditions. Since the study of the regulatory mechanism of LEA gene expression is an important feature of stress molecular biology, a detailed discussion on structure, function and expression regulation of LEA proteins in higher plants is presented in this chapter.

Chapter 3 – Abiotic stresses, such as drought, high salinity, extreme temperatures, heavy metals and oxidative stress affect plant growth and extremely decrease crop productivity.

Plants respond to these environmental challenges through the activation and the regulation of specific stress related genes. The dehydration responsive element binding (DREB) transcription factors, that strongly up-regulates many downstream genes, play an important role in plant environmental stress tolerance, thereby increase efficiency of plant production.

Recently, several research approaches were developed to understand the molecular mechanisms of stress responses based on the investigation of the involvement of such transcription factors on signaling pathways related to abiotic stress tolerance.

The present review discusses the functions of the DREB transcription factors in plant abiotic stress responses and summarizes the recent advances in elucidating stress-response mechanisms and their biotechnological applications. This review examines also the progress of the genetic engineering approaches developed with these DREB transcription factors in the main crops and model plants in order to elucidate their impact on plant tolerance.

In: Abiotic Stress
Editor: Annabella Ferro

ISBN: 978-1-63117-622-7
© 2014 Nova Science Publishers, Inc.

*Chapter 1*

# THE PIVOTAL ROLE PLAYED BY TRANSCRIPTION FACTORS, SMALL RNAS AND EPIGENETIC MODIFICATIONS IN THE REGULATION OF ABIOTIC STRESS RESPONSES IN PLANTS

*Paola A. Palmeros-Suárez and John P. Délano-Frier**
Centro de Investigación y de Estudios Avanzados del I.P.N.
Unidad Irapuato, Km 9.6 del Libramiento Norte Carretera Irapuato-León,
Apartado Irapuato, Gto., México

## ABSTRACT

Abiotic stresses have become one the principal factors limiting crop productivity world-wide. Abiotic stresses like salinity, drought, high and low temperatures, ozone, heavy metals, and low concentration of nutrients in degraded soils, cause deleterious effects in almost all phenological plant stages, from germination to full plant development. It is known that plants respond readily when challenged by abiotic stresses through the modified regulation of many genes. This is a manifold response consisting of transcriptional and translational mechanisms that allow the adaptation of plants to many unfavorable environments.

---

* Corresponding author: jdelano@ira.cinvestav.mx.

To date, many genetic and molecular strategies have been developed to identify those factors that are important for plants to adapt to stressful environments. As a result, a comprehensive knowledge regarding the pattern of induction of many genes during adverse conditions, have uncovered a complex network of inter-connected responses. At a further level of regulation, the function of transcription factors has been found to be essential to regulate the proper activation of an adequate answer to given environmental stimuli consisting of specific posttranslational and epigenetic modifications, including changes in nucleosome distribution, histone modification, DNA methylation, and npcRNAs (non-protein-coding RNAs) that play an important role in abiotic stress gene networks as well.

In the post-genomics era, data garnered from comprehensive analyses using increasingly efficient technologies such as transcriptomics, proteomics and metabolomics, have increased our understanding of the complex regulatory networks associated with stress adaptation and tolerance; besides, natural variations for stress responses have been observed at different levels of integration using strategies like classical linkage and association (LD) mapping.

Moreover, classical breeding approaches have revealed that stress tolerance traits are dispersed in various quantitative trait loci (QTLs), a factor which hinders their rapid genetic selection. Geneticists are working on fundamental mechanisms of adaptation which could be useful to crop breeders looking for natural variants that could optimize environmental resources and provide targets for breeding programs.

Taking together all of the above aspects, this chapter describes some of the most recent investigations about molecular, physiological and biochemical changes that are triggered in response to abiotic stress in plants. Its main message is that in order to deal with changing and challenging environmental conditions, plants exhibit a wide range of integrated responses which usually display complex quantitative variation.

# 1. INTRODUCTION

Stress may be defined as any negative effect or condition that a living organism may suffer. Among the most common abiotic stresses are those related to drought, salinity, extreme temperatures (in both the low or high ranges), heavy metal accumulation and mineral nutrient deficiencies; these are known to impose severe production constrains on food, fodder, fiber and fuel production [1]. Food security of many countries depends upon irrigated agriculture; besides, the increasing population demands more production from the shrinking availability of agricultural land. This situation raises the imperative to use crops able to withstand stressful environmental conditions, and in the same time, increase yields, in order to face an ever growing global demand. Genetic enhancement of abiotic stress tolerance in strategic crops is one of the important strategies implemented to enhance their productivity in these circumstances [1].

It has been observed that in the last years the combined average temperatures over global land and ocean surfaces have increased considerably. The national climatic data center, published that the average temperature in November 2013 was the record highest for the 134-year period of record, at 0.78°C (1.40°F) above the 20th century average of 12.9°C (55.2°F). And the combined global land and ocean average surface temperature for the year was 0.62°C (1.12°F) above the 20th century average of 14.0°C (57.2°F), tying with 2002 as the fourth warmest such period on record (http://www.ncdc.noaa.gov/sotc/global/2013/11/). Therefore, it has become imperative to find viable solutions, not only to increase crop yields for a growing population, but to generate crops capable of tolerating the consequences of global warming.

According to the nature, timing and intensity of the stress episode, tolerance can be defined either as the ability of plants to survive severe stresses and complete their cycle, or to achieve acceptable yields when subjected to milder degrees of stress [2]. The response level depends on the species, development stage, and the metabolic state of the plant, as well as the duration and intensity of the stress [3].

The series of events leading to a response to stress begins with the perception of a specific signal. This signal activates a specific group of internal responses that lead to changes in gene expression and/ or in metabolism [4, 5, 6]. These changes represent the effort made by the plants to overcome the stress situation, maintain homeostasis and, eventually, adapt [7, 4, 8].

One of the most critical stresses affecting yield crops is drought, which can occur at different stages of the plant's development, with different effects on plant function. This necessarily requires distinct mechanisms for tolerance; besides, a variety of additional abiotic stresses commonly occur during drought, such as high temperatures, elevated concentrations of salts and other toxic solutes and low availability of nutrients; these can vary in time and space [9]. The responses of plants to drought stress are highly complex, involving adaptive changes. Early responses of plants to drought stress usually help the plant to survive for a relatively short time, while the long-term acclimation of the plant subjected to drought is characterized by the accumulation of a certain group of metabolites associated with structural molecular capabilities designed to improve plant functioning under drought stress [10]. Survival in hostile environments involves developing mechanisms of tolerance, resistance, or avoidance. Plants that develop tolerance to a given factor can, over time, overcome the effects of this factor without injury. Conversely, resistance to a given stressful environment requires the deployment of a set of counter measures, while the acceleration of the plant life cycle to allow flowering before a drought period is a well-known avoidance mechanism [8].

There is a diversity of mechanisms which can be used by plants to tolerate a diversity of environmental stresses [9]. Plants are able to make adjustments in morphology, phenology, physiology and/ or biochemistry in response to stress. This usually involves the up-regulation of various genes, the expression of which can contribute to mitigate the effect of stress and lead to the adjustment of the cellular milieu and, eventually, to plant tolerance [11]. Physiological responses involve the accumulation of stress-associated proteins and metabolites [12, 13]. Genetic responses can be mediated through epigenetic regulation, including RNA-directed DNA methylation, histone and DNA modifications, which play an important role in altering gene expression against abiotic stress [14]. Other genetic approaches, including transcriptomic, proteomic, and molecular studies have identified many stress-related genes, which are generally classified into two major groups. The first one is involved in signaling cascades, transcriptional control, and the degradation of transcripts or proteins, whereas members of the second group function in membrane protection, osmoprotection, as antioxidants and as reactive oxygen species (ROS) scavengers. Simultaneously, metabolic pathways are adjusted to regain homeostasis in a changing environment [15].

Many studies in this direction have been implemented in an attempt to identify stress-regulated genes. Some have shown that plants that are exposed to different stresses, employ genes that are regulated in singular ways, but that nevertheless induce similar defense responses. Probably this is because drought, salinity, extreme temperatures, and oxidative stress are interconnected, and may induce similar effects on plants. Salinity and drought, for example, cause resembling disturbances in plant cells, including membrane and protein damage and disruption of ion distribution [8].

Here, we will describe the progress that has been achieved so far in the genetic engineering of plants, which will focus on the overexpression of regulatory genes. This strategy has proven to be more accurate than conventional or molecular breeding, by allowing the overexpression and/ or introduction into the genome of highly selective master genes. In addition, the role of other regulatory molecules, such as small RNA and plant hormones, will be explored. Finally, we will briefly discuss the genetic breeding techniques which are already employed as experimental strategies to breed better crops. Irrespective of the methodology employed, the principal objective in plant breeding is to obtain plants that combine higher yields, reliable yield stability, better quality, and obvious stress-resistant characters (abiotic and biotic) over different years and locations [8]. Classical breeding approaches have revealed that stress tolerance traits are dispersed in various quantitative trait loci (QTLs), which makes genetic selection of these traits difficult [15]. However, the identification and use of molecular markers and introgression of genomic portions (QTLs) involved in stress tolerance still represents an attractive alternative, although undesirable agronomic characteristics from the donor plants may be introduced into the target plant [8].

## 2. GENETIC REGULATION BY TRANSCRIPTION FACTORS

Transcription factors (TFs) play critical roles in controlling intrinsic developmental processes and responses to external stimuli by influencing the expression of downstream target genes; they have also been confirmed to improve drought resistance in transgenic plants [16, 17, 18].

TFs interact with cis-acting elements present in the promoter region of various stress-responsive genes and thus activate cascades or whole networks of genes that act together to enhance simultaneous tolerance towards multiple stresses. This property greatly enhances the attractiveness of TFs as regulatory genes for manipulation of abiotic stress tolerance [19]. Thus, stress responsive TFs are powerful tools for genetic engineering, as their overexpression can lead to either the up- or down-regulation of a whole array of genes placed under their control [19]. Several studies have revealed that the principal TFs involved in abiotic stress responsive pathways are distributed mainly in the AP2/ERF, bZIP, NAC, MYB, C2H2 zinc finger, and WRKY families [20, 21, 22, 23, 24, 25]. In the next section, we will describe the salient results obtained by diverse research groups, all which have underlined the important participation of the TFs in plant stress responses.

## 2.1. Ethylene Responsive Factor (ERF)

The role of the ERF family of TFs in the regulation of abiotic stress responses has been well-documented in several plant species. The AP2/ERF family conforms a large group of plant-specific TFs that share a well-conserved DNA-binding domain that includes the AP2, RAV, DREB, ERF and other subfamilies. These TFs been implicated in sugar, hormone, and redox signaling in the context of abiotic stresses [26]. As mentioned above, this TF family includes the dehydration-responsive element-binding (DREB) protein/ C-repeat binding factors (CBFs) that recognize the DRE/ CRT cis-element in order to regulate the expression of stress-responsive genes [27]. The expression of DREB1/ CBF genes under different abiotic stress situations has been investigated extensively in a genetically wide variety of plant species. Many studies have reported that the expression of most of these genes generally displays a tendency to decrease sharply after a rapid initial increase, as exemplified by the achievement of high and transient levels of expression, reached between 1 and 4 h under cold stress treatment, in various plant species [19]. For instance, the DREB1 subgroup of Arabidopsis consists of six members. Among these, the importance of the DREB1A/ CBF3, DREB1B/ CBF1 and DREB1C/ CBF2 TFs, which are rapidly induced in response to cold stress, has been well established. Moreover, transgenic Arabidopsis constitutively overexpressing any one of these three DREB1s/ CBFs not only display significantly improved tolerance to freezing, but also to drought and high salinity, whereas the suppression of DREB1A/CBF3 or DREB1B/CBF1 causes a reduction in freezing tolerance [27]. Thus, genetic engineering has proven to be a powerful tool to explore the effect of DREB1/CBF and may be employed to deploy the DREB/CBF TFs for increasing the abiotic stress tolerance potential of crop plants. Not surprisingly, transcriptome analyses of DREB1/CBF-overexpressing Arabidopsis have revealed that many cold-inducible genes are up-regulated in these plants. Likewise, metabolomic analyses have demonstrated that cold stress induces the accumulation of many metabolites, such as sugars, a large proportion of which also accumulate in transgenic Arabidopsis plants constitutively overexpressing DREB1s/CBFs. Cold-inducible DREB1/CBF genes have also been isolated from numerous plant species, such as tomato, oilseed rape, wheat, rye, rice and maize [27]. Furthermore, the overexpression of Arabidopsis CBF genes in transgenic *Brassica napus* plants induces the expression of orthologs of Arabidopsis CBF-targeted genes and increases the freezing tolerance of important cereal crops, such as wheat and rye, which are known to cold acclimate, as well as in tomato, a freezing-sensitive plant that does not cold acclimate [28].

All these findings support that TFs function in a similar way in abiotic stress tolerance among dicots and monocots. However, in most cases, transgenic plants showed severe growth retardation under normal growth conditions [19]. The growth retardation in transgenic tomato overexpressing DREB1B was prevented by exogenous application of gibberellic acid (GA), which suggests that the hyper-accumulation of CBF1 protein may interfere with GA biosynthesis [29]. Delayed flowering is the second most common problem in plants overexpressing a DREB1 gene; thus, both dwarfism and delayed flowering are thought to be related to interference in GA metabolism [30, 31]

Most of the DREB2 genes that have been reported are responsive to water stress or heat shock, but DREB2 genes from grass species are also responsive to cold stress [32, 33, 34]. In Arabidopsis, DREB2A and DREB2B are induced by dehydration, high salinity and heat [35, 36], while DREB2C shows a retarded induction by heat that occurs later than that of DREB2A or DREB2B [23]. DREB2C, DREB2D and DREB2F are also high salinity-inducible and DREB2E is ABA-inducible, though its expression is rather weak [26]. Therefore, DREB2A and DREB2B seem to be the major DREB2 TFs involved in dehydration-inducible gene expression through DRE/ CRT in an ABA-independent pathway [26, 36]. In contrast, to DREB1s/ CBF TFs, ectopic overexpression of DREB2A does not cause significant changes in plant growth, gene expression or stress tolerance, which suggests posttranslational regulation of the DREB2A protein [37, 27].

A study performed by Zhu et al. [38], reported that three ERF TFs identified in resistant Chinese wild *Vitis pseudoreticulata*, namely *VpERF1, VpERF2*, and *VpERF3*, were involved in abiotic stress-responsive pathways. These findings were based on experiments in which young grapevine seedlings were challenged by cold (4°C), heat (44°C), and drought stress. The data produced showed that *VpERF1* was significantly induced by drought and heat, but remained insensitive to cold stress, while *VpERF2* was strongly induced by cold, heat, and drought. *VpERF3* was also induced by cold and heat but at low levels relative to *VpERF2*. These results indicated that the three *VpERFs* may contribute differently to alleviate the negative effects of abiotic stress response [38].

## 2.2. Basic Leucine Zipper (bZIP)

The (bZIP) TF family is one of the largest TF families in plants and its members have diverse roles, particularly in plant stress-responses and hormone signal transduction [39, 40, 41]. To activate downstream gene expression, the bZIP TFs interact with ABA-responsive elements (ABREs). These are cis-acting elements containing the PyACGTGGC core sequence and are present in the promoter region of ABA-inducible genes. Hence, bZIP TFs are designated as ABRE-binding factors (ABFs) or ABRE-binding proteins (AREBs) [42, 43, 44]. It is known that bZIP ABF genes, such as ABF2/ AREB1, ABF4/ AREB2, and ABF3, are up-regulated by ABA, dehydration, and salinity stress in vegetative tissues in Arabidopsis [39, 44]. Also, the constitutive overexpression of ABF3 in Arabidopsis and rice results in enhanced drought tolerance [46]. Moreover, in rice, the overexpression of the positive regulators of ABA signaling, OsbZIP23 and OsbZIP72, enhances abiotic stress tolerance [24, 47]. In drought, salinity and oxidative stress (induced by methyl viologen application) treated seedlings, *OsABF2* expression was highly induced within 3 h, whereas when seedlings were subjected to cold conditions (4°C), *OsABF2* expression was similarly induced within 3 h and slightly decreased after 24 h [48]. In the same way, *ABF1* to *4* and *OsbZIP23*, *OsbZIP72*, *OsABI5* and *OsABF1* were also strongly induced by drought, salt, cold, and ABA treatments and their characterization has confirmed the critical roles they play in ABA dependent signaling and stress tolerance [45, 49, 24, 47, 50].

## 2.3. MYB Proteins

So far, MYB proteins have been shown to be involved in many significant physiological and biochemical processes, including regulation of primary and secondary metabolism, control of cell development and the cell cycle, participation in defense and in response to various biotic and abiotic stresses, in addition to flavonoid and hormone biosynthesis, and signal transduction [51]. Extensive studies in various plant species have provided a better understanding of the MYB gene family. For instance, Chen et al. [51] tested the expression pattern of the 30 *MYB* genes in peanut roots under salt stress tress, which revealed that the expression of four *R2R3-MYB* genes (*AhMYB1*, *AhMYB2*, *AhMYB6* and *AhMYB7*) was enhanced under this stressful condition. The expression of *3R-MYB* and the other five *R2R3-MYB* genes underwent no obvious change, while the expression of 20 MYB-related proteins showed different accumulation patterns in roots under this stress condition. The transcript abundance of *AhMYB12*, *AhMYB18*, *AhMYB28* and *AhMYB30* also increased under salt stress, while the expression of *AhMYB11*, *AhMYB15* and *AhMYB17* decreased. The other 13 peanut *MYB*-related genes showed no obvious change in expression under salt stress. These findings indicated that at least eight peanut *MYB* genes were induced by salt stress, namely *AhMYB1*, *AhMYB2*, *AhMYB6*, *AhMYB7*, *AhMYB12*, *AhMYB18*, *AhMYB28* and *AhMYB30* [51]. On the other hand, the latter eight salt-induced genes showed different expression patterns in response to drought and cold stresses when their expression was monitored within a 72 h time framework. Furthermore, the analysis of the relationship between the abiotic stress-induced *MYB* genes and ABA revealed that these eight genes were also expressed in peanut roots and/ or leaves treated with exogenous ABA. All genes, except *AhMYB12* and *AhMYB18*, were induced by ABA in roots, with transcript accumulation occurring mostly within the first 6 h after starting the stress treatment, thereby suggesting that MYB family proteins have complex functions in the regulation of abiotic stress in peanut, and most probably other plants [51].


## 2.4. WRKY Proteins

Members of *WRKY* TF family are characterized by the presence of one or two highly conserved WRKY domains containing an N-terminal WRKYGQK motif and a C terminal zinc finger motif [52]. Both conserved domain elements are necessary for high binding affinity of WRKY proteins to the Wbox (C/T)TGAC(C/T) [52, 53], which is often present in the promoters of many defense-related genes and *WRKY* genes themselves [54, 55, 56]. Several Arabidopsis *WRKY* (*AtWRKY*) genes respond to high salinity, cold, and osmotic stress [57]. Besides, overexpression of *AtWRKY25* and *AtWRKY33* enhanced salinity tolerance in Arabidopsis [58]. Likewise, the overexpression of wheat *TaWRKY2* and *TaWRKY19* WRKY TFs was shown to regulate drought stress tolerance in Arabidopsis by activating downstream target genes related to drought resistance, such as *DREB*, *RD29*, and *COR6.6* [59]. However, a study performed in rice by Peng, et al., [60], showed that the expression of *OsWRKY82* was induced only by mechanical wounding and heat shock treatment, and no induction was observed in cold, high salinity, and dehydration conditions.

## 2.5. Nuclear Factor Y Family

Recently, the Nuclear Factor Y family (NF-Y) of TFs, which bind specifically to the CCAAT box, was identified in drought responsive pathways and also identified as a CCAAT-box binding factor (CBF) [61, 62, 63]. Recent reports have elucidated some functions of the NF-Y TF family. These include a role of NF-YB and NF-YC members in Arabidopsis, tobacco, and wheat in the regulation of light responses, including flowering time [61, 64, 65]. In addition, *AtNF-YB6* (*L1L*) and *AtNF-YB9* (*LEC1*) are involved in embryo development in seeds [66, 67], whereas NF-Y family members in plants have been found to function in drought stress. [62]. In this respect, it was observed that transgenic plants overexpressing *AtNF-YA5* displayed reduced stomatal aperture and leaf water loss and significantly promoted drought resistance. Moreover, *AtNF-YA5* was found to be regulated transcriptionally by ABA and post-transcriptionally by miR169, while the overexpression of *AtNF-YB1* conferred improved performance in Arabidopsis under drought treatments, with transformed plants showing higher water potential and photosynthesis rates. In addition, transgenic maize plants transformed with *ZmNF-YB2*, an homologue of *AtNF-YB1*, showed a drought tolerance response based on increased chlorophyll content, stomatal conductance, and photosynthesis rates, as well as decreased leaf temperature under drought conditions, all of which were deemed to contribute to the grain yield advantage obtained [61].

## 2.6. Cys2/His2-Type Zinc-finger Proteins

Recent investigations concerned with the role of TFs in response to stress have revealed their participation in both biotic and biotic stress, involving an active cross talk between different kinds of adverse environmental conditions. A relevant example is the *Di19* (for *Drought-induced*) Arabidopsis gene, which is a member of a gene family encoding seven Cys2/His2-type zinc-finger proteins, most of which have unknown functions. Nevertheless, it is known that *AtDi19* functions as a transcriptional regulator that is involved in Arabidopsis' responses to drought stress, which surprisingly coincides with the up-regulation of *PR1*, *PR2*, and *PR5* pathogenesis-related protein genes. These results are in agreement with the observation that the *Di19T*-DNA insertion mutant *di19* was much more sensitive to drought stress, whereas the *Di19*-overexpressing line proved to be more tolerant to drought stress compared with wild type plants [68].

## 2.7. Homeodomain-Leucine Zipper TF

Another well-studied family of TFs is one composed of homeodomain-leucine zipper (HD-Zip) proteins that have important participation in numerous physiological processes in plants. In a study realized by Song et al., [69] a homeobox-leucine zipper gene from *Medicago truncatula*, designated *MtHB2*, was identified while monitoring the expression profile of *M. truncatula* plants exposed to low temperatures. To evaluate the role of *MtHB2* in response to abiotic stresses, they generated transgenic Arabidopsis plants capable of overexpressing the *MtHB2* gene, which were subsequently exposed to diverse of abiotic stresses. Contrary to expectations, the *MtHB2*-overexpressing transgenic plants generated were more sensitive to drought, salt and freezing stress than wild-type plants. It was found that the expression of *MtHB2* in Arabidopsis resulted in transgenic plants that accumulated lower amounts of proline and soluble sugars, but had increased levels of malondialdehyde (MDA) and $H_2O_2$ than their wild-type counterparts challenged with the same abiotic stresses. The reduced accumulation of proline and soluble sugars was thought to be responsible for the lower osmolality detected in the transgenic plants and to contribute to the reduced osmo-regulating efficiency of the transgenic plants. On the other hand, it was suggested that their higher levels of MDA and $H_2O_2$ made them more susceptible to oxidative damage under the abiotic stress conditions imposed. These findings demonstrated that *MtHB2* encodes a novel stress-responsive HD TF that may play a negative role in the regulation of abiotic stress response mechanisms in plants [69].

## 2.8. NAC Proteins

NAC TFs have a variety of important functions, not only in plant development, but also in abiotic stress responses. For instance, transgenic Arabidopsis and rice plants overexpressing stress-responsive *NAC* (*SNAC*) genes exhibited improved drought tolerance. These studies indicate that SNAC factors have important roles in the control of abiotic stress tolerance and that their overexpression can improve stress tolerance via biotechnological approaches.

**Table 1. Examples of important transcription factors induced in response to abiotic stress conditions in plants**

| Transcription Factor | Gene | Plant | Stress Condition | Reference |
|---|---|---|---|---|
| CBF | *DREB1A/CBF3, DREB1B/CBF1, DREB1C/CBF2* | Arabidopsis | Cold | [19, 28, 46] |
| | *DREB1* | Aloe | Cold | [19] |
| | *LpCBF3* | Rye grass | Cold | [19] |
| | *PeDREB1* | *Phyllostachys edulis* | Cold | [19] |
| | *Ptcbfb* | *Poncirus trifoliata* | Cold | [19] |
| | *OsAP211* | Rice | Cold | [19] |
| | *EguCBFs* | Eucalyptus | Cold | [19] |
| | Bn115 | Canola | Cold | [19] |
| DREB1/CBFs | *AtDREB1A* | Arabidopsis | Drought, salt | [35, 36] |
| | *BrCBF* | Chinese cabbage | Cold, salt, drought, ABA | [19] |
| | *MbDREB1* | Apple | Cold, salt, drought, ABA | [19] |
| | *CrCBF* | *Catharanthus roseus* | Cold | [19] |
| | *OsDREB2A, OsDREB2s* | Rice | Heat, drought, salt | [32] |
| | *OsDREB1C, OsDREB1F* | | Cold, salt, drought, ABA | [19] |
| | *VviDREB1* | *Vaccinium vitisidaea* | Cold, salt, drought, ABA | [19] |
| | *TaDREB1* | *Triticum aestivum* L | Cold, salt, drought | [33] |
| | *Wdreb2* | Wheat | cold, drought, salt, ABA | [34] |
| | DBF1 | Maize | Drought, salt, ABA | [34] |
| | *HvDRF1.1 , HvDRF1.2 HvDRF1.3* | Barley | ABA, drought, salt | [34] |
| ERF | *VpERF1, VpERF2, VpERF3* | *Vitis* pseudoreticulata | cold, heat, drought | [38] |
| bZIP | *ABF1, ABF2, ABF3, ABF4* | Arabidopsis | Salt, cold, drought, ABA | [45] |
| | *OsABI5* | Rice | ABA, salt | [50] |
| | *OsbZIP72* | | Drought, salt, cold, ABA | [47] |
| | *OsABF2* | | Drought, salinity, cold, oxidative stress, ABA. | [48] |
| | *bZIP* | Common bean, Tepary bean | Drought | [41] |

| Transcription Factor | Gene | Plant | Stress Condition | Reference |
|---|---|---|---|---|
| MYB | *AhMYB1, AhMYB2, AhMYB6, AhMYB7, AhMYB12, AhMYB18, AhMYB28, AhMYB30* | Peanut | Salt | [51] |
| | *PeSCL7* | Poplar | Drought, NaCl | [127] |
| | *TaMYB33* | Wheat | Drought, ABA, NaCl | [128] |
| | *ScMYBAS1* | Sugarcane | Drought, salt | [129] |
| | *StMYB1R-1* | Potato | Drought, ABA, NaCl | [130] |
| WRKY | *WRKY25, WRKY33** | Arabidopsis | NaCl | [58] |
| | *OsWRKY82** | Rice | Cold, salt, drought | [60] |
| | *BhWRKY1* | *Boea hygrometrica* | Drought | [25] |
| | *TaWRKY2, TaWRKY19* | *Wheat* | drought, salt | [59] |
| | *Hv-WRKY38* | Barley | Cold, drought | [131] |
| NFY | *AtNF-YB, NFYA5* | Arabidopsis | Drought | [61, 62] |
| | *Zm NF-YB2* | Maize | Drought | [61] |
| | *1TaNF-YA, 5 TaNF-YB, 3 TaNF-YC , TaDr1* | Wheat | Drought | [64] |
| | *PdNF-YB7* | Poplar | Drought | [18] |
| | *NF-YC2* | Arabidopsis | Oxidative stress | [65] |
| | *NF-YC* | Tobacco | Oxidative stress | [65] |
| Cys2/His2 Zinc-finger | *Di19** | Arabidopsis | Drought | [68] |
| | *ThZF1* | Salt cress | Salt, drought | [134] |
| | *AtAZF2, AtSTZ* | Arabidopsis | Salt, drought | [133] |
| | *SCOF-1* | Soybean | Cold | [132] |
| HD-Zip | *MtHB2* | *Medicago truncatula* | Cold, salt, PEG | [69] |
| | *AtHB7, AtHB12, HAT2,HAT22* | Arabidopsis | Drought | [135] |
| | *CpHB1, CpHB2* | *Craterostigma plantagineum* | Drought | [135] |
| | *HaHB4* | Sunflower | Drought | [135] |
| NAC | *OsNAC6, ONAC045* | Rice | Cold, drought, salt, ABA | [70] |
| | *SsNAC23** | Sugarcane | Cold, drought | [70] |
| | *ATAF1* | Arabidopsis | Drought | [23] |
| | *GmNAC* | Soybean | Drought | [70] |

*Also induced by biotic stress.

Although these TFs can bind to the same core NAC recognition sequence, recent studies have demonstrated that the effects of NAC factors on growth are different. Moreover, the NAC proteins are capable of functioning as homo- or heterodimers. Thus, SNAC factors can be useful for improving stress tolerance in transgenic plants, although the mechanism(s) responsible for mediating the stress tolerance of these homologous factors in plants appears to be complex and remains largely unknown. Recent studies also suggest that these TFs might be able to regulate a signaling cross-talk between stress responses and plant growth [70].

The studies mentioned above, describe specific examples that illustrate the important participation of TFs in the adaptive responses that plants deploy to cope with different types of stress. They further corroborate the need to gain further understanding of their targets and mode of action. Additional information about these crucial stress-responsive TFs is shown in Table 1.

## 3. THE ROLE OF SMALL RNAs IN ABIOTIC STRESS RESPONSES IN PLANTS

RNA regulatory processes that control transcription, degradation and stabilization are major mechanisms that determine the levels of mRNAs in plants [71]. Transcriptional and post-transcriptional regulations of RNAs are drastically altered during plant stress responses. As a result of these molecular processes, plants are capable of adjusting to changing environmental conditions [71].

Understanding gene regulation mechanisms is important for genetic improvement of abiotic stress resistance of crops. Recent studies in the area of RNA regulation have increased our understanding of how plants respond to environmental stresses. Various types of RNA regulatory factors and processes such as small RNAs, antisense RNAs, alternative splicing, RNA decay, RNA stability and RNA-binding proteins have emerged as new research areas involved in plant stress responses [71].

Recently, transcriptome analyses using high-density microarrays and high throughput sequencing technologies have revealed a vast number of non-coding RNAs (ncRNAs) that are expressed from un annotated genomic regions as non-protein coding small RNAs [1, 71]. Among them, micro RNAs (miRNAs) and small interfering RNAs (siRNAs) are found to be very important riboregulators in plants. Various types of sRNAs differ in their mode of biogenesis and in their gene regulation function. sRNAs are involved in gene regulation at both transcriptional and post-transcriptional levels. They are known to regulate growth and development of plants [72]. Furthermore, sRNAs especially plant miRNAs, have been found to be involved in various stress responses, such as oxidative stress, mineral nutrient deficiency, dehydration, and even mechanical stimulus [73, 72]. Small non-coding regulatory RNAs are about 20–24 nucleotides in length; they are single stranded RNAs processed from an endogenous transcript that contains a local hairpin structure by the Dicer-like (DCl) family of plant enzymes [74]. Endogenous small RNAs can be divided in microRNAs (miRNAs) and short interfering RNAs (siRNAs), both of which are small noncoding RNAs that have recently emerged as important regulators of mRNA degradation, translational repression, and chromatin modification [75]. miRNAs are known to play important regulatory roles in plants by targeting mRNAs for cleavage or translational repression [76].

It has been suggested that the regulation of stress-related genes and miRNAs are somehow correlated [75]. In this context, vast studies that have been performed in plant model systems such as in *A. thaliana* [75], *Oryza sativa* [77], *Zea mays* [78], *Nicotiana tabacum* [79], *Populus trichocarpa* [80], *Gossypium hirsutum* [81] *B. napus* [82] *and Triticum aestivum* [83], have corroborated this assumption. Thus, the majority of miRNA target genes have been found to encode various TFs or important functional enzymes, and to play important roles in plant development and in response to various abiotic stresses [72]. The miRNAs identified are either up- or down-regulated in response to stress treatments that impact plant growth and developmental processes [72].

Almost a decade ago, Sunkar and Zhu [75], identified novel abiotic stress-regulated miRNAs and siRNAs, by constructing a library of small RNAs from Arabidopsis seedlings exposed to dehydration, salinity, cold stress or ABA. They observed that some of the miRNAs were preferentially expressed in specific tissues, and several were either up-regulated (e.g., miR393, miR397b and miR402) or down-regulated (e.g., miR389a) by abiotic stresses. Shortly thereafter, Zhao et al. [84], confirmed that miR-169g was induced by drought, and suggested that the induction, which was more prominent in roots than in shoots, could be regulated directly by DRE binding TFs. Another work, performed in *Phaseolus vulgaris*, showed a moderate but clear increase in the accumulation of miRS1, miR1514a, and miR2119 in response to drought treatment, while a greater accumulation of miR159.2, miR393, and miR2118 was detected in response to the same treatment. Besides, they observed increased accumulation of miRS1 and miR159.2 in response to NaCl addition [85].

It has also been observed that target genes of abiotic stress-regulated miRNAs encode proteins for ubiquitin mediated proteolysis (E3 ubiquitin ligases, SCF complex components and F-box proteins). This is highly relevant considering that protein degradation plays a crucial role in various plant abiotic stress responses. For instance, many abiotic stresses induce senescence, which more often than not occurs in older leaves in order to supply the nutrients, mainly rich in N, which are required by young leaves and reproductive organs. Besides, leaf senescence is also used as a strategy to reduce water loss under drought stress. Also, heat stress is well-known to cause the denaturation of proteins, which in subsequent events are either repaired or undergo ubiquitin-mediated proteolysis. In fact, heat stress has been shown to increase conjugated ubiquitination in plants [86, 87], whereas ubiquitination may also regulate gene expression under cold stress [88].

Other target genes that are regulated by miRNAs are those known to be involved in sugar sensing (At4g03190, identical to *Glucose Repression Resistance 1, GRR1*); this is in accordance with findings showing that abiotic stresses have a strong influence on the sugar status of cells due to a decrease in photosynthesis, change in metabolism, and demand for high maintenance respiration [1].

The accumulation of reactive oxygen species (ROS) as a result of various environmental stresses is a major cause of crop productivity loss [89, 90]. ROS affect many cellular functions by damaging nucleic acids, oxidizing proteins, and causing lipid peroxidation. Stress-induced ROS accumulation is counteracted by intrinsic antioxidant systems in plants that include a variety of enzymatic scavengers, such as superoxide dismutases (SOD), some of which employ copper-zinc as metal co-factors (Cu/Zn-SOD), which are localized in different cellular compartments [91]. Accordingly, the overexpression of a cytosolic and chloroplastic Cu/Zn-SODs from pea (the first one recognized as a homolog of the *CSD2* present in *A. thaliana*) in transgenic tobacco plants increased ozone tolerance [92], and led to an augmented tolerance to high light and low temperature stresses [93], respectively. A related study [94], reported that the expression of miR398 was down-regulated by conditions conducive to oxidative stress, such as high light, heavy metals and methyl viologen application. This study also found that the observed down-regulation was important for the post-transcriptional induction of the Cu/Zn *CSD1* and *CSD2* SOD genes. Furthermore, they showed that relieving miRNA-directed suppression by overexpression of a miR398-resistant version of CSD2 led to significant improvement of plant resistance to oxidative stress conditions. These findings provided solid evidence that suppressing the expression of a miRNA is important for plant adaptation to abiotic stresses.

Research done by Borsani, et al. [95], revealed that the antisense overlapping gene pair consisting of the D1-pyrroline-5-carboxylate dehydrogenase (*P5CDH*) and *SRO5* genes, generates two types of siRNA. The first gene codes for an enzyme that functions as an intermediate in proline synthesis and catabolism, and has a key role in stress responses and accumulation of ROS. P5CDH can also promote apoptosis which may be mediated, at least partly, through ROS accumulation. The second gene (*SRO5*) is an unknown function gene whose expression is induced by salt. A 24-nt siRNA (referred as 24-nt *SRO5-P5CDH* nat-siRNA) was detected in NaCl-treated adult Arabidopsis plants and seedlings and its induction was found to be required to initiate siRNA formation, thereby representing a new mechanism of *P5CDH* regulation mediated by endogenous siRNAs under stress conditions. Thus, the *SRO5-P5CDH* nat-siRNAs together with the P5CDH and SRO5 proteins are key components of a regulatory loop controlling ROS production and stress responses. The nat-siRNA-mediated cross regulation of *P5CDH* and *SRO5* mRNAs and the functional relationship of these two proteins suggest a regulatory model that may be applied to other cis-antisense gene pairs [95].

Similar investigations by Zhu et al. [96], led to the patent of novel mechanisms that control plant gene expression in response to stress. A relevant finding was the discovery that genes encoding laccase-like proteins (LAC) and a regulatory subunit of casein kinase (CKB3) were down-regulated by salt stress in Arabidopsis. Furthermore, this down-regulation was found to be caused by a salt stress-induced transcriptional up-regulation of miR397, which directed the cleavage of *LAC* and *CKB3* transcripts. Interestingly, the overexpression of miR397 in transgenic Arabidopsis enhanced *LAC* and *CKB3* transcript cleavage as well as its tolerance to salt stress. These results demonstrated that down-regulation of *LAC* and *CKB3* transcripts, guided by miR397, is essential for salt tolerance. Hence, manipulation of the expression of such miRNAs can be an effective new approach for the improvement of the plant salt tolerance [72, 96].

Another example of miRNAs that regulate the plant's response to salinity stress involves the miR169 family. Zhao et al. [97], showed that miR169g and miR169n,o were significantly induced during salt stress, which led to the selective cleavage of a CCAAT-box recognized by the NF-YA TF. Besides, the existence of a cis-acting ABA responsive element (ABRE) in the upstream region of miR-169n,o suggested that miR-169n,o may be regulated by ABA [97].

Many miRNAs detected in maize are also induced by salt stress and directly regulate TFs involved in plant development and organ formation. Ding et al. (2009) observed that Myb, NAC1 and HD-ZIP TFs are the targets of zma-miR159a/b, zma-miR164a/b/c/d and zma-miR166l/m, respectively. Other predicted miRNA TF targets, such as MADS-box proteins and zinc-finger proteins, have been reported as salt stress-responsive factors in plants. Several other miRNA targets, include auxin response factor 8, ethylene-responsive element-binding factor and GAMyb, all of which are involved in phytohormone signaling cascades. Additional miRNA targets encode proteins that play important roles in diverse metabolic pathways or are involved in various physiological processes [72, 78].

This same study, revealed that the cis-elements of five of the miRNA gene promoters included ABA-responsive elements (ABRE), MYB binding sites (MBS), salicylic acid-responsive elements (e.g., having a TCA motif), heat shock-responsive elements (HSE), low temperature-responsive elements (LTR), GC-motifs, anaerobic-response elements (ARE) and gibberellin-responsive elements (GARE). Interestingly, each of the salt stress-responsive miRNAs was found to have more than one stress-responsive cis-acting element in their promoter regions, indicating that these miRNAs are involved in the regulation of signaling cascades associated to various biotic and abiotic stress responses. The three most frequently appearing cis-acting motifs of these salt stress-responsive miRNA genes were MBS, ARE and ABRE. In addition, miRNA members of the same miRNA family often share similar cis-acting motifs in their promoter regions, suggesting that the miRNA members belonging to a certain miRNA family may respond to the same biotic, abiotic or phytohormone stimuli [78]. Further experimentation by this group led to the cloning of miRNAs belonging to miR396 family, which were reported to be down-regulated in the presence of salt stress, in contrast to miR474 and miR395, which were induced by salt stress, and thereby suppressed the expression of their respective targets [78].

An additional study by Jung and Kang [98] reported that miR417 was regulated by ABA, dehydration and salt stress in Arabidopsis. High salt stress was shown to moderately decrease transcript levels of miR417, while ABA treatment and dehydration stress resulted in an initial up-regulation followed by its subsequent down-regulation. Moreover, seeds of Arabidopsis plants overexpressing miR417, showed a decreased rate of seed germination and a lower seedling survival rate, as compared to wild-type plants, when subjected to salt and ABA treatment. These findings were indicative that miRNA417 plays a role as a negative regulator of seed germination in Arabidopsis plants under salt stress conditions [98].

Another miRNA found to be up-regulated by various abiotic stresses or stress associated hormones, like ABA, dehydration, cold and high NaCl is miR393. This miRNA has been reported to regulate the expression of transcripts encoding the F-box auxin receptor and the transport inhibitor response 1 (TIR1) protein. TIR1, in turn, targets AUX/IAA proteins for proteolysis by SCF-E3 ubiquitin ligases in an auxin-dependent manner. These proteins are known to be necessary for various auxin-induced growth and developmental processes [72, 99, 100]. These observations supplied the first documented evidence that different abiotic stresses lead to increased TIR1 mRNA degradation or translational repression via miR393 accumulation, which negatively impacts auxin signaling and seedling growth.

Studies carried out in wheat, have revealed that the expression of four siRNA varied greatly in seedlings subjected to various stress treatments such as cold (4°C for 2 h), heat (40°C for 2 h), high salt (200 mM NaCl) and water deprivation. Thus, experimental evidence showed that both siRNA 002061_0636_3054.1 and siRNA 005047_0654_1904.1 were down-regulated by heat, NaCl and dehydration stress, while the latter was also up-regulated by cold stress. In addition, siRNA 080621_1340_0098.1 was up-regulated by cold and down- regulated by heat but not by NaCl and dehydration stress, while siRNA 007927_0100_2975.1 was down-regulated by cold, NaCl and dehydration stress but not by heat stress. The study concluded that the genes targeted by these siRNAs could play key roles in regulating stress responses in wheat [83]

It is important to emphasize that miRNAs, TFs, hormones and other functional molecules are involved in the orchestration of a complete plant response to adverse environmental conditions. An illustrative example was described by Reyes [101] who showed that in germinating Arabidopsis seeds, miR159 accumulates in response to ABA and drought during seed germination in order to negatively regulate ABA-dependent responses during seed germination. Besides, this worker found that ABA-induced miR159 accumulation requires ABI3 and other known components of the ABA signaling pathway. His results indicated that ABA induced accumulation of miR159 is part of a homeostatic mechanism designed to direct MYB33 and MYB101 transcript degradation in order to desensitize hormone signaling during seedling stress responses.

The past few years have witnessed an explosive increase in research reports on plant miRNAs, as evidenced by the identification of more than 700 miRNAs in a variety of plant species. Large numbers of miRNA targets were predicted, some of which were validated or confirmed experimentally. These pioneer works will provide a foundation for future research in the field. An understanding of miRNA mediated gene regulation could lead to novel strategies for improving important plant traits, such as stress tolerance [74]. However, a word of caution must be introduced at this point to indicate that, in spite of a large body of experimental evidence suggesting that stress regulation may imply a potential function of regulated miRNA in stress responses, it is still not obviously clear that these regulatory molecules are involved in stress adaptation responses [102]. The studies of abiotic stress-associated plant miRNAs in the last few years have focused on miRNA discovery, especially at the whole genome scale. We now have a long list of abiotic stress-associated miRNAs whose expression is altered under different stresses and in a variety of plants. These efforts have strongly suggested that miRNAs play important roles in stress responses and have built a solid foundation for future research. Whereas genome-wide expression profiling will likely to continue in the near future, there is an urgent need for detailed functional characterizations of individual miRNAs [103]. Finally, the most recent investigations about the regulation of miRNA have focused in the TF binding motifs (TFBMs) that are located in the upstream region of the target genes, which regulate their TF-associated expression.

## Table 2. Participation of representative small RNAs in abiotic stress responses in plants

| miRNA | Plant | Stress Condition | Reference |
|---|---|---|---|
| miR393, miR397b, miR402 | Arabidopsis | Cold, drought, NaCl, ABA | [75] |
| miR389a | | Cold | [75] |
| | | Drought, NaCl, ABA | |
| miR169 family | | Drought | [97] |
| miRNA417 | | ABA, drought, salt | [98] |
| miRNA159 | | ABA, drought | [101] |
| miR1514a, miR2119, miR393, miR2118 | Common bean | Drought | [85] |
| miRS1 and miR159.2 | | Salt , drought | [85] |
| siRNA002061_0636_3054.1 | Wheat | Cold | [83] |
| siRNA 005047_0654_1904.1 | | Heat, NaCl, drought | |
| 080621_1340_0098.1 | | Cold  Heat | [83] |
| siRNA 007927_0100_2975 | | Heat | [83] |
| | | Cold, NaCl, drought | |
| miR159a/b,miR164a/b/c/d | Maize | Salt | [78] |
| miR166l/m | | | [72] |
| miR474 miR395 | | | |

Up-regulated. Down-regulated.

The architecture of TFBMs may be conserved in a particular class of genes or in genes co-expressed in response to a particular stress. This phenomenon may also operate in the regulation of miRNA genes. In this respect, Devi et al. [104] performed a study in rice designed to identify and determine the positional conservation of over-represented, unique or well-known *cis* motifs present in abiotic stress-related miRNAs. After selecting 136 miRNAs that strongly suggested an involvement in abiotic stress regulation, they found that miRNAs grouped together in a given cluster were not always involved in the regulation of the same kind of stress, thereby suggesting the hypothesis that miRNAs are independent transcription units that are finely regulated according to different environmental cues [104].

Understanding the roles of small RNAs in transcriptome homeostasis, cellular tolerance, and phenological and developmental plasticity of plants under abiotic stress and recovery will certainly help genetic engineering approaches to enhance abiotic stress resistance in crop plants [1]. An informative summary of the most relevant small RNAs described in this section is shown in Table 2.

# 4. OTHER TECHNOLOGIES USED FOR CROP IMPROVEMENT

In the post-genomics era, comprehensive analyses using functional genomics technologies such as transcriptomics, proteomics, and metabolomics have significantly increased our understanding of the complex regulatory networks associated with stress adaptation and tolerance [105].

Transcriptome analysis technologies have advanced to the point where high-through-put DNA sequencers and high density microarrays such as tiling arrays are readily available. These technologies provide new opportunities to analyze non-coding RNAs and also to clarify aspects of epigenetic regulation of gene expression. Such analyses have markedly increased our understanding of global plant systems in responses and adaptation to stress conditions [105].

Genetic regulation and epigenetic regulation, including changes in nucleosome distribution, histone modification, DNA methylation, and npcRNAs (non-protein-coding RNA) play important roles in abiotic stress gene networks. [105]. Recent advanced technologies, such as genome-wide transcriptome analysis, have revealed that a vast amount of non-coding RNAs (ncRNAs), apart from the well-known housekeeping ncRNAs such as rRNAs, tRNAs, small nuclear RNAs (snRNAs) and small nucleolar RNAs (snoRNAs), are expressed under abiotic stress conditions. The various ncRNAs are involved in chromatin regulation, modulation of RNA stability and translational repression during abiotic stress responses [106]. ncRNAs are transcribed from intergenic regions, antisense strands of protein-coding genes and also from pseudogenes [106]. In 2008, Matsui et al. [107], estimated that approximately 80% of previously unannotated up-regulated transcripts arise from antisense strands of sense transcripts. There was a significant linear correlation between the expression ratios (stress-treated/ untreated) of the sense transcripts with those of the antisense transcripts. Recently obtained data have revealed a complex interaction between transcriptional regulators which function to fine-tune responses to various environmental stimuli [106].

On another hand, it has been observed that changes of transcriptional states in response to environmental stresses are coupled with chromatin remodeling, which are accompanied by post translational modification of histone N-tails, such as acetylation, methylation, and phosphorylation. It is noteworthy to mention that various modifications of histones occur during stress responses in plants [108]. Additionally, the results published by Sokol et al. [109], indicated that quiescent tobacco and Arabidopsis cell lines showed a nucleosomal response to abiotic stresses and ABA treatment and may represent suitable models for the study of chromatin-mediated mechanisms of stress tolerance in plants.

Recently, plant epigenetics has gained interest, not only as a subject of basic research but also as a possible new source of beneficial traits for plant breeding [110]. This is supported by the findings obtained from the study of mechanisms responsible of the formation of heritable epigenetic gene variants (epialleles) and also of transposon mobility regulation, both of which have suggested that they represent subjects that could be exploited to broaden plant phenotypic and genetic variation. Research efforts in this direction could improve long-term plant adaptation to environmental challenges and increase productivity [110].

Epigenetics is defined as a change in gene expression without base sequence alteration. An epigenetically acquired trait within an organism is transmissible from cell to cell, and is commonly observed during ontogeny [111]. It is known that epigenetic mechanisms involving dynamic changes in chromatin properties and the biogenesis of small RNAs contribute to transcriptional and post-transcriptional regulation of gene expression important for stress responses that result, not only in short-term acclimation, but also in long-lasting and heritable phenotypic traits. The likelihood exists that plants may be able to perceive stresses during vegetative growth and then proceed to 'memorize' them, possibly by epigenetic mechanisms, in cell lineages that later contribute to the germline [110]. Environmentally induced epigenetic states could thus be passed to the progeny. Besides, germline progenitor cells in particular need to be protected against such environmental effects to maintain the integrity of their genomes and epigenomes and thus ensure the continuity of transgenerational inheritance [110].

As mentioned above, the induction of alternative epigenetic states not only triggers the formation of novel epialleles but also promotes the movement of DNA transposons and retroelements that are very abundant in plant genomes. Retrotransposons, which proliferate by means of reverse transcription of RNA intermediates, comprise a major portion of plant genomes. Plants often change their genome size and organization during evolution by rapid proliferation and deletion of long terminal repeat (LTR) retrotransposons [112]. In plants and mammals, retrotransposons are transcriptionally silenced by DNA methylation, which in Arabidopsisis is propagated at CG dinucleotides [113]. However, it has been observed that a *copia-type* Évadé retrotransposon (EVD) evaded suppression of its movement during inbreeding of hybrid epigenomes consisting of *met1*-methyltransferase mutants and wild-type-derived chromosomes. A genetic testing of host control of the EVD life cycle showed that transcriptional suppression occurred by CG methylation supported by RNA-directed DNA methylation [113]. On transcriptional reactivation, subsequent steps of the EVD cycle were inhibited by different elements, and genome re-sequencing demonstrated retrotransposition of EVD, but of no other potentially active retroelements, when this combination of epigenetic mechanisms was compromised. This study convincingly demonstrated that epigenetic control of retrotransposons extends beyond transcriptional suppression and can be individualized for particular elements [113]. Therefore, novel 'epigenetically induced' heritable phenotypes can be of true epigenetic or genetic nature and both types can increase the ability of plants to adapt to environmental challenges. Such scenario suggests that the way in which epigenetic regulation could contribute to the enrichment of novel traits related to plant stress adaptation, either directly or indirectly, could proceed through the controlled generation and exploitation of retrotransposon-induced genetic diversity [110].

As already mentioned, DNA methylation is importantly involved in gene silencing; this often involves the hyper-methylation of promoter sequences [114]. It is thought that an environmental stimulus can induce adaptive responses through heritable chromatin modifications mediated by DNA methylation. Thus, DNA methylation appears to satisfy conditions for epigenetic inheritance, being reversibly changeable over generations, inducing no base alteration and strongly affecting gene expression through regulation of chromatin structure [115].

An example of the participation of DNA methylation in response to adverse stimuli was reported by Akimoto et al. [115] in rice plants. Their report presented evidence that genes constantly silenced due to hyper-methylation can be transcriptionally activated by a de-methylation process, which is often accompanied by the conferral of new phenotypes. They also found that changes in both methylation and phenotype are stably inherited. In this respect, they showed that de-methylation activated a disease resistance gene that activated a corresponding trait, and that both hypo-methylation and the acquired phenotype associated with this condition, were stably inherited by the progeny.

A notable feature of this study was that the methylation/ de-methylation patterns are faithfully transmitted to the offspring, thereby suggesting that methylation patterns in somatic cells are maintained in germ cells. In consequence, the manipulation of a mechanism such as the one involving the methylation machinery could exploited to regulate gene expression by flexibly tuning their methylation status. Such strategy may potentially allow plants to gain or lose particular heritable traits, which could be maintained by controlling the methylation patterns of the corresponding genes [115]. In this respect, Steward et al. [116] reported that levels of DNA methylation change in the nucleosome cores in maize root tissues in response to cold stress or environmental cues. Such behavior allowed these workers to speculate that DNA methylation may function as a common switch of gene expression and that naturally induced changes in DNA methylation may result in heritable epigenetic modification of gene expression. Moreover, Boyko and Kovalchuk [117] mentioned that the plasticity of plant phenotypes cannot be simply explained by genetic changes such as point mutations, deletions, insertions and gross chromosomal rearrangements. They argued that probably other lasting mechanisms of adaptation, such as those controlling heritability by reversible epigenetic modifications capable of regulating gene expression without changing DNA sequence, could be a viable alternative to cope with environmental stresses through several plant generations, [117]. A finding that was in agreement with this proposal was reported by Molinier, et al. [118] who showed that treatment of *A. thaliana* plants with short-wavelength ultraviolet-C radiation or flagellin, an elicitor of plant defense responses, increased the frequency of somatic homologous recombination of a transgenic reporter which persisted in the subsequent, untreated, generations. Besides, the epigenetic trait of enhanced homologous recombination could be transmitted through both the maternal and paternal crossing partner, and proved to be dominant. These results prompted the conclusion that environmental factors lead to increased genomic flexibility even in successive, untreated generations, and may increase the potential for adaptation.

Another common form of genome natural diversity that has begun to be studied in plants is gene copy number variation (CNV), which may have broad implications for model organism research, evolutionary biology, and crop science [119]. It has been reported that genome rearrangements can rapidly generate genetic diversity by causing CNV, affecting contiguous sets of genes that lead to genetic and phenotypic variability among individuals in the population [117]. A study performed by DeBolt showed that biotic and abiotic stresses could trigger CNVs. A propensity for duplication and non-repetitive CNVs was documented at 28°C, which correlated with the highest stress level in Arabidopsis plants and permitted to infer a potential CNV–environmental interaction. A broad diversity of gene species were affected by CNVs, supporting a model whereby segmental CNV, together with the genes encoded within these regions, contribute to the adaptive capacity of plants through natural genome variation [119]. However, it is important to consider that a more complex genome with multiple chromosomes may have a greater tendency to adapt to stress or may be less able to acquire a recessive phenotype. Differences are also expected between plants with short and long generation periods (ranging from a few months to many years) and between sexually and vegetatively propagated plants [117].

A new approach has emerged that dissects yield and integrative traits that influence stress tolerance into heritable traits by using phenotyping platforms with model-assisted methods [2]. Tailoring genotypes with acceptable performance under adverse environmental conditions is essential for the sustainability of crop production under the threat of climate change. Although conventional breeding has been successful in some cases, it still requires time-consuming hard work because it involves a plethora of loci (mostly QTLs) whose effects differ in the many possible stress scenarios. Furthermore, field trials have revealed that a given QTL exerts, in the most favorable cases, a significant effect on yield in only half of the tested environments encompassing a broad range of different climatic scenarios [120, 121].

Alternative strategies involving the use of natural genetic diversity under diverse environmental conditions require a much larger number of candidate lines and of environmental scenarios [2]. However, the development of new techniques allowing a much more efficient performance of physiological and genetic analyses has enabled the study of large networks of field experiments. A representative example of this generation of novel analytical tools is high-throughput phenotyping, which allows reproducible measurement of drought-related physiological traits in hundreds of genotypes, both in controlled conditions and in the field [122]. Besides, methods have been developed that enable the mathematical description of traits that vary rapidly with environmental conditions, via model assisted phenotyping and meta-analyses of large datasets [123]. It has been observed that modeling may well provide the most efficient way to find optimal parameters, particularly when the effect on yield of a given trait depends on a trade-off between positive and negative effects, with different optima depending on the prevailing environmental conditions. The multiplicity of stressful environments, which are difficult to handle experimentally, can now be addressed in robust models. Thus, if a model adequately describes the effects of genetic variability in a few climatic scenarios, it can then be extrapolated to a much larger number of scenarios in order to evaluate the comparative advantage of a given allele in different environments [124].

Based on several studies, it seems plausible to assume that any trait involved in stress tolerance can be considered as having positive, negative or no effect on yield depending on the stress scenario. It follows that the genetic variability of these traits should be analyzed and dissected independently of its consequences on yield in a given environment [2].

Finally, the identification of loci (genes or QTLs) governing variability of traits involved in stress tolerance is of greatest interest if their effects can be scaled up to crop performance in the field under different stress scenario [2, 125]. Thus, the improvement of crop yield has been possible through the indirect manipulation of quantitative trait loci (QTLs) that control heritable variability of the traits and physiological mechanisms that determine biomass production and its partitioning.

The co-location between QTLs simple traits and yield essentially applies to constitutive traits with a large effect on yield. The assumption is that the concurrent effects of each QTL on yield and the target trait are controlled by the same locus and not by tight linkage. In durum wheat, for example, field testing has uncovered two QTLs that consistently influence plant height, kernel weight and yield, across a broad range of soil moisture environments. Hence, the fine mapping and eventually cloning of these QTLs will be facilitated by measuring plant height rather than yield. Once the cloning of these two QTLs is completed, a better understanding of the mode of action and agronomic value of the allelic series present at these loci will be possible [2, 126].

# CONCLUSION

The improvement of crop resistance to abiotic stress will depend on many factors, including the future discovery of promising genes and a significant advancement towards a better understanding of stress protection pathways. As we have seen in this chapter, TFs represent pivotal players capable of coordinating the regulation of many other genes that leads to a complete response able to cope with different types of stresses. In addition, other emerging mechanisms involving the participation of small RNAs and epigenetic modifications have gained importance concomitantly with the growing knowledge of their active participation in stress response regulation in plants.

Ideally, the application of this knowledge will require the utilization of efficient gene engineering strategies, applicable to a wider range of species and having no negative effects on crop yield, plant development and food safety, which are based on a holistic knowledge of the principal components of a desired plant stress-resistance trait.

# REFERENCES


[1]   Chinnusamy, V., Zhu, J., Zhou, T., Zhu, J.-K. (2007). Small Rnas: big role in abiotic stress tolerance of plants. In M. Jenks, P. Hasegawa, & S. M. Jain (Eds.), *Advances in Molecular Breeding Toward Drought and Salt Tolerant Crops SE - 10* (pp. 223–260). Springer Netherlands. doi:10.1007/978-1-4020-5578-2_10

[2]   Tardieu, F., Tuberosa, R. (2010). Dissection and modelling of abiotic stress tolerance in plants. *Current Opinion in Plant Biology, 13* (2), 206–12. doi:10.1016/j.pbi.2009.12.012

[3]   Reddy, A., Chaitanya, K., Vivekanandan, M. (2004). Drought-induced responses of photosynthesis and antioxidant metabolism in higher plants. *Journal of Plant Physiology, 161*(11), 1189–1202. doi:10.1016/ j.jplph.2004.01.013

[4]     Hazen, S., Wu, Y., Kreps, J. (2003). Gene expression profiling of plant responses to abiotic stress. *Functional & Integrative Genomics*, *3*(3), 105–111. doi:10.1007/s10142-003-0088-4

[5]     Shao, H.-B., Guo, Q.-J., Chu, L.-Y., Zhao, X.-N., Su, Z.-L., Hu, Y.-C., Cheng, J.-F. (2007). Understanding molecular mechanism of higher plant plasticity under abiotic stress. *Colloids and Surfaces. B, Biointerfaces*, *54* (1), 37–45. doi:10.1016/j.colsurfb.2006.07.002

[6]     Agrawal, G. K., Jwa, N.-S., Lebrun, M.-H., Job, D., Rakwal, R. (2010). Plant secretome: unlocking secrets of the secreted proteins. *Proteomics*, *10*(4), 799–827. doi:10.1002/pmic.200900514

[7]     Altman, A. (2003). From plant tissue culture to biotechnology: Scientific revolutions, abiotic stress tolerance, and forestry. *In Vitro Cellular & Developmental Biology - Plant*, *39* (2), 75–84. doi:10.1079/IVP2002379

[8]     Macedo, A. F. (2012). Abiotic stress responses in plants, 41–61. *Abiotic Stress Responses in Plants: Metabolism, Productivity and Sustainability* doi:10.1007/978-1-4614-0634-1

[9]     Roy, S. J., Tucker, E. J., Tester, M. (2011). Genetic analysis of abiotic stress tolerance in crops. *Current Opinion in Plant Biology*, *14* (3), 232–9. doi:10.1016/j.pbi.2011.03.002

[10]   Pinhero, R. G., Rao, M. V., Paliyath, G., Murr, D. P., Fletcher, R. (1997). Changes in activities of antioxidant enzymes and their relationship to genetic and paclobutrazol-induced chilling tolerance of maize seedlings. *Plant Physiology*, *114*(2), 695–704. Retrieved from http://www.pubmedcentral.nih.gov/articlerender.fcgi?artid=158354&tool=pmcentrez&rendertype=abstract

[11]   Mahajan, S., Tuteja, N. (2005). Cold, salinity and drought stresses: an overview. *Archives of Biochemistry and Biophysics*, *444* (2), 139–158. doi:10.1016/j.abb.2005.10.018

[12]   Cushman, J. C., Bohnert, H. J. (2000). Genomic approaches to plant stress tolerance. *Current Opinion in Plant Biology*, *3* (2), 117–24. Retrieved from http://www.ncbi.nlm.nih.gov/pubmed/10712956

[13]   Vinocur, B., Altman, A. (2005). Recent advances in engineering plant tolerance to abiotic stress: achievements and limitations. *Current Opinion in Biotechnology*, *16* (2), 123–132. doi:10.1016/j.copbio.2005.02.001

[14]   Chinnusamy, V., Zhu, J.-K. (2009). Epigenetic regulation of stress responses in plants. *Current Opinion in Plant Biology*, *12* (2), 133–9. doi:10.1016/j.pbi.2008.12.006

[15] Pardo, J. M. (2010). Biotechnology of water and salinity stress tolerance. *Current Opinion in Biotechnology*, *21* (2), 185–196. doi:10.1016/j.copbio.2010.02.005

[16] Shinozaki, K., Yamaguchi-Shinozaki, K., Seki, M. (2003). Regulatory network of gene expression in the drought and cold stress responses. *Current Opinion in Plant Biology*, *6* (5), 410–417. doi:10.1016/S1369-5266(03)00092-X

[17] Valliyodan, B., Nguyen, H. T. (2006). Understanding regulatory networks and engineering for enhanced drought tolerance in plants. *Current Opinion in Plant Biology*, *9* (2), 189–95. doi:10.1016/j.pbi.2006.01.019

[18] Han, X., Tang, S., An, Y., Zheng, D.-C., Xia, X.-L., Yin, W.-L. (2013). Overexpression of the poplar NF-YB7 transcription factor confers drought tolerance and improves water-use efficiency in Arabidopsis . *Journal of Experimental Botany*, *64* (14), 4589–4601. doi:10.1093/jxb/ert262

[19] Akhtar, M., Jaiswal, A., Taj, G., Jaiswal, J. P., Qureshi, M. I., Singh, N. K. (2012). DREB1 / CBF transcription factors: their structure , function and role in abiotic stress tolerance in plants. *Journal of Genetics*, *91* (3), 385–395.

[20] Abe, H., Yamaguchi-shinozaki, K., Urao, T., Hosokawa, C. D. (1997). Role of Arabidopsis MYC and MYB homologs in drought- and abscisic acid-regulated gene expression, *The Plant Cell*, *9*, 1859–1868.

[21] Kizis, D., Lumbreras, V. (2001). Role of AP2 / EREBP transcription factors in gene regulation during abiotic stress. *Federation of European Bochequimal Societies Letters*, *498*, 187–189.

[22] Sakamoto, H., Maruyama, K., Sakuma, Y., Meshi, T., Iwabuchi, M. (2004). Arabidopsis Cys2 / His2-Type zinc-finger proteins function as transcription repressors under drought, *136* (1) 2734–2746. doi:10.1104/pp.104.046599.2734

[23] Lu, P.-L., Chen, N.-Z., An, R., Su, Z., Qi, B.-S., Ren, F., Chen, J.,Wang, X.-C. (2007). A novel drought-inducible gene, ATAF1, encodes a NAC family protein that negatively regulates the expression of stress-responsive genes in Arabidopsis. *Plant Molecular Biology*, *63* (2), 289–305. doi:10.1007/s11103-006-9089-8

[24] Xiang, Y., Tang, N., Du, H., Ye, H., Xiong, L. (2008). Characterization of OsbZIP23 as a key player of the basic leucine zipper transcription factor family for conferring abscisic acid sensitivity and salinity and

drought tolerance in rice. *Plant Physiology, 148* (4), 1938–52. doi:10.1104/pp.108.128199

[25] Wang, Z., Zhu, Y., Wang, L., Liu, X., Liu, Y., Phillips, J., Deng, X. (2009). A WRKY transcription factor participates in dehydration tolerance in *Boea hygrometrica* by binding to the W-box elements of the galactinol synthase (BhGolS1) promoter. *Planta, 230* (6), 1155–66. doi:10.1007/s00425-009-1014-3

[26] Sakuma, Y., Liu, Q., Dubouzet, J. G., Abe, H., Shinozaki, K., Yamaguchi-Shinozaki, K. (2002). DNA-binding specificity of the ERF/AP2 domain of Arabidopsis DREBs, transcription factors involved in dehydration- and cold-inducible gene expression. *Biochemical and Biophysical Research Communications, 290* (3), 998–1009. doi:10.1006/bbrc.2001.6299

[27] Mizoi, J., Shinozaki, K., Yamaguchi-Shinozaki, K. (2012). AP2/ERF family transcription factors in plant abiotic stress responses. *Biochimica et Biophysica Acta, 1819* (2), 86–96. doi:10.1016/j.bbagrm.2011.08.004

[28] Jaglo, K. R., Kleff, S., Amundsen, K. L., Zhang, X., Haake, V., Zhang, J. Z., Deits, T., Thomashow, M. F. (2001). Components of the Arabidopsis C-repeat / dehydration- responsive element binding factor cold-response pathway are conserved in *Brassica napus* and other plant species. *Plant Physiology, 127*, 910-917. doi:10.1104/pp.010548.910

[29] Hsieh, T., Lee, J., Charng, Y., Chan, M. (2002). Tomato plants ectopically expressing Arabidopsis CBF1 show enhanced resistance to water deficit stress 1(*130*), 618–626. doi:10.1104/pp.006783.KIN1

[30] Achard, P., Gong, F., Cheminant, S., Alioua, M., Hedden, P., Genschik, P. (2008). The cold-inducible CBF1 factor-dependent signaling pathway modulates the accumulation of the growth-repressing DELLA proteins via its effect on gibberellin metabolism. *The Plant cell, 20* (8), 2117–29. doi:10.1105/tpc.108.058941

[31] Magome, H., Yamaguchi, S., Hanada, A., Kamiya, Y., Oda, K. (2008). The DDF1 transcriptional activator upregulates expression of a gibberellin-deactivating gene, GA2ox7, under high-salinity stress in Arabidopsis. *The Plant journal: for Cell and Molecular Biology, 56* (4), 613–26. doi:10.1111/j.1365-313X.2008.03627.x

[32] Matsukura, S., Mizoi, J., Yoshida, T., Todaka, D., Ito, Y., Maruyama, K., Shinozaki, K., Yamaguchi-Shinozaki, K. (2010). Comprehensive analysis of rice DREB2-type genes that encode transcription factors involved in the expression of abiotic stress-responsive genes. *Molecular*

*Genetics and Genomics, 283* (2), 185–96. doi:10.1007/s00438-009-0506-y

[33] Shen, Y.-G., Zhang, W.-K., He, S.-J., Zhang, J.-S., Liu, Q., Chen, S.-Y. (2003). An EREBP/AP2-type protein in *Triticum aestivum* was a DRE-binding transcription factor induced by cold, dehydration and ABA stress. *TAG. Theoretical and Applied Genetics. Theoretische und angewandte Genetik, 106*(5), 923–30. doi:10.1007/s00122-002-1131-x

[34] Egawa, C., Kobayashi, F., Ishibashi, M., Nakamura, T., Nakamura, C., Takumi, S. (2006). Differential regulation of transcript accumulation and alternative splicing of a DREB2 homolog under abiotic stress conditions in common wheat. *Genes & Genetic Systems, 81* (2), 77–91. Retrieved from http://www.ncbi.nlm.nih.gov/pubmed/16755132

[35] Liu, Q., Kasuga, M., Sakuma, Y., Abe, H., Miura, S., Yamaguchi-Shinozaki, K., Shinozaki, K. (1998). Two transcription factors, DREB1 and DREB2, with an EREBP/AP2 DNA binding domain separate two cellular signal transduction pathways in drought- and low-temperature-responsive gene expression, respectively, in Arabidopsis. *The Plant Cell, 10* (8), 1391–406. Retrieved from http://www.pubmedcentral.nih.gov/articlerender.fcgi?artid=144379&tool=pmcentrez&rendertype=abstract

[36] Nakashima, K., Shinwari, Z. K., Sakuma, Y., Seki, M., Miura, S., Shinozaki, K., Yamaguchi-Shinozaki, K. (2000). Organization and expression of two Arabidopsis DREB2 genes encoding DRE-binding proteins involved in dehydration- and high-salinity-responsive gene expression. *Plant Molecular Biology, 42* (4), 657–65. Retrieved from http://www.ncbi.nlm.nih.gov/pubmed/10809011

[37] Sakuma, Y., Maruyama, K., Qin, F., Osakabe, Y., Shinozaki, K., Yamaguchi-Shinozaki, K. (2006). Dual function of an Arabidopsis transcription factor DREB2A in water-stress-responsive and heat-stress-responsive gene expression. *Proceedings of the National Academy of Sciences of the United States of America, 103* (49), 18822–7. doi:10.1073/pnas.0605639103

[38] Zhu, Z., Shi, J., Xu, W., Li, H., He, M., Xu, Y., Xu, T., Yang, Y., Cao, J., Wang, Y. (2013). Three ERF transcription factors from Chinese wild grapevine *Vitis pseudoreticulata* participate in different biotic and abiotic stress-responsive pathways. *Journal of Plant Physiology, 170* (10), 923–33. doi:10.1016/j.jplph.2013.01.017

[39] Jakoby, M., Weisshaar, B., Dröge-Laser, W., Vicente-Carbajosa, J., Tiedemann, J., Kroj, T., Parcy, F. (2002). bZIP transcription factors in

Arabidopsis. *Trends in Plant Science, 7*(3), 106–111. doi:10.1016/S1360-1385(01)02223-3

[40] Uno, Y., Furihata, T., Abe, H., Yoshida, R., Shinozaki, K., Yamaguchi-Shinozaki, K. (2000). Arabidopsis basic leucine zipper transcription factors involved in an abscisic acid-dependent signal transduction pathway under drought and high-salinity conditions. *Proceedings of the National Academy of Sciences of the United States of America, 97* (21), 11632–7. doi:10.1073/pnas.190309197

[41] Rodriguez-Uribe, L., O'Connell, M. A. (2006). A root-specific bZIP transcription factor is responsive to water deficit stress in tepary bean (*Phaseolus acutifolius*) and common bean (*P. vulgaris*). *Journal of experimental botany, 57* (6), 1391–8. doi:10.1093/jxb/erj118

[42] Niu, X., Renshaw-Gegg, L., Miller, L., Guiltinan, M. (1999). Bipartite determinants of DNA-binding specificity of plant basic leucine zipper proteins. *Plant Molecular Biology, 41* (1), 1–13. doi:10.1023/A:1006206011502

[43] Kim, S.Y., Chung H.-J., Thomas, T.L. (1997). Isolation of a novel class of bZIP transcription factors that interact with ABA-responsive and embryo-specification elements in the Dc3 promoter using a modified yeast one-hybrid system. *The Plant Journal 11 (6),* 1237-1251.

[44] Yamaguchi-Shinozaki, K., Shinozaki, K. (2005). Organization of cis-acting regulatory elements in osmotic- and cold-stress-responsive promoters. *Trends in Plant Science, 10* (2),.elsevier.com/retrieve/pii/S1360138504002985

[45] Choi, H. -i. (2000). ABFs, a family of ABA-responsive element binding factors. *Journal of Biological Chemistry, 275* (3), 1723–1730. doi:10.1074/jbc.275.3.1723

[46] Oh, S., Song, S. I., Kim, Y. S., Jang, H., Kim, S. Y., Kim, M., Kim, Y-K., Nahn B.H., Kim J-K (2005). Arabidopsis CBF3 / DREB1A and ABF3 in transgenic rice increased tolerance to abiotic stress without stunting growth, *Plant Physiology, 138*, 341–351. doi:10.1104/pp.104.059147.1

[47] Lu, G., Gao, C., Zheng, X., Han, B. (2009). Identification of OsbZIP72 as a positive regulator of ABA response and drought tolerance in rice. *Planta, 229* (3), 605–15. doi:10.1007/s00425-008-0857-3

[48] Hossain, M. A., Cho, J.-I., Han, M., Ahn, C.-H., Jeon, J.-S., An, G., Park, P. B. (2010). The ABRE-binding bZIP transcription factor OsABF2 is a positive regulator of abiotic stress and ABA signaling in

rice. *Journal of plant physiology*, *167* (17), 1512–20. doi:10.1016/j.jplph.2010.05.008

[49] Zhu, J.-K. (2002). Salt and drought stress signal transdution in Plants, *Annual Review of Plant Biology*, *53,* 247-273 doi:10.1146/annurev.arplant.53.091401.143329.

[50] Zou, M., Guan, Y., Ren, H., Zhang, F., Chen, F. (2008). A bZIP transcription factor, OsABI5, is involved in rice fertility and stress tolerance. *Plant molecular biology*, *66* (6), 675–83. doi:10.1007/s11103-008-9298-4

[51] Chen, N., Yang, Q., Pan, L., Chi, X., Chen, M., Hu, D., Yang, Z., Wang, D., Wang, M., Yu, S. (2014). Identification of 30 MYB transcription factor genes and analysis of their expression during abiotic stress in peanut (*Arachis hypogaea* L.). *Gene*, *533* (1), 332–45. doi:10.1016/j.gene.2013.08.092

[52] Eulgem, T., Rushton, P. J., Robatzek, S., Somssich, I. E. (2000). The WRKY superfamily of plant transcription factors. *Trends in plant science*, *5* (5), 199-206. http://linkinghub. elsevier.com/retrieve/pii/S1360138500016009

[53] Maeo K, S Hayashi, H Kojima-Suzuki, A. M., Morikami, A., Nakamura, K. (2001). Role of conserved residues of the WRKY domain in the DNA-binding of tobacco WRKY family proteins. *Bioscience Biotechnology and Biochemistry, 65 (11)* 2428-2436.

[54] Maleck, K., Levine, A., Eulgem, T., Morgan, A., Schmid, J., Lawton, K. A., Dang, L., Dietrich, R. A. (2001). The transcriptome of *Arabidopsis thaliana* during systemic acquired resistance. *Nature, 26*, 403-410.

[55] Yu, D., Chen, C., Chen, Z. (2001). Evidence for an important role of WRKY DNA binding proteins in the regulation of NPR1 gene expression. *The Plant Cell, 13,* 1527–1539.

[56] Turck, F., Zhou, A., Somssich, I. E. (2004). Stimulus-dependent , romoter-specific binding of transcription factor WRKY1 to its native promoter and the defense-related gene *PcPR1-1* in Parsley, *The Plant Cell, 16*, 2573–2585. doi:10.1105/tpc.104.024810.1

[57] Seki, M., Narusaka, M., Ishida, J., Nanjo, T., Fujita, M., Oono, Y., Kamiya, A., Nakajima, M., Enju, A., Sakurai, T., Satou, M., Akiyama, K., Taji, T., Yamaguchi-Shinozaki, K., Carninci, P., Kawai, J., Hayashizaki, Y., Shinozaki, K. (2002). Monitoring the expression profiles of 7000 Arabidopsis genes under drought, cold and high-salinity stresses using a full-length cDNA microarray. *The Plant Journal, 31* (3), 279–292. doi:10.1046/j.1365-313X.2002.01359.x

[58] Jiang, Y., Deyholos, M. (2009). Functional characterization of Arabidopsis NaCl-inducible WRKY25 and WRKY33 transcription factors in abiotic stresses. *Plant Molecular Biology, 69* (1-2), 91–105. doi:10.1007/s11103-008-9408-3

[59] Niu, C.-F., Wei, W. E. I., Zhou, Q.-Y., Tian, A.-G., Hao, Y.-J., Zhang, W.-K., Ma, B., Lin., Q., Zhang, Z-B., Zhan, J-B., Chen, S.-Y. (2012). Wheat WRKY genes TaWRKY2 and TaWRKY19 regulate abiotic stress tolerance in transgenic Arabidopsis plants. *Plant, Cell & Environment, 35* (6), 1156–1170. doi:10.1111/j.1365-3040.2012.02480.x

[60] Peng, X., Tang, X., Zhou, P., Hu, Y., Deng, X., He, Y., Wang, H. (2011). Isolation and expression patterns of rice WRKY82 transcription factor gene responsive to both biotic and abiotic stresses. *Agricultural Sciences in China, 10* (6), 893–901. doi:10.1016/S1671-2927(11) 60074-6

[61] Nelson, D. E., Repetti, P. P., Adams, T. R., Creelman, R. A., Wu, J., Warner, D. C., Anstrom, D.C., Bensen, R.J., Castiglioni P.P.,. Donnarummo, M. G., Hinchey, B. S., Kumimoto, R.W., Maszle, R.D., Canales, R.D., Krolikowski, K.A., Dotson, S.B., Gutterson, N., Ratcliffe O. J.,Heard, J. E. (2007). Plant nuclear factor Y (NF-Y) B subunits confer drought tolerance and lead to improved corn yields on water-limited acres. *Proceedings of the National Academy of Sciences, 104 (42),* 16450–16455.

[62] Li, W-X., Oono, Y., Zhu, J., He, X-J., Wu, J-M., Iida, K., Lu, X-Y., Cui, X., Jin, H., Zhu, J. (2008). The Arabidopsis NFYA5 transcription factor is regulated transcriptionally and posttranscriptionally to promote drought resistance, *The Plant Cell, 20,* 2238–2251. doi:10.1105/ tpc.108.059444

[63] Stephenson, T., McIntyre, C. L., Collet, C., Xue, G.-P. (2011). TaNF-YB3 is involved in the regulation of photosynthesis genes in *Triticum aestivum. Functional & Integrative Genomics, 11*(2), 327–340. doi:10.1007/s10142-011-0212-9

[64] Stephenson, T. J., McIntyre, C. L., Collet, C., Xue, G.-P. (2007). Genome-wide identification and expression analysis of the NF-Y family of transcription factors in *Triticum aestivum. Plant Molecular Biology, 65,* 77–92. doi:10.1007/s11103-007-9200-9

[65] Hackenberg, D., Keetman, U., Grimm, B. (2012). Homologous NF-YC2 subunit from Arabidopsis and Tobacco is activated by photooxidative stress and induces flowering. *International journal of molecular sciences, 13* (3), 3458–77. doi:10.3390/ijms13033458

[66] Kwong, R. W., Bui, A. Q., Lee, H., Kwong, L. W., Fischer, R. L., Goldberg, R. B., Harada, J. J. (2003). LEAFY COTYLEDON1-LIKE Defines a class of regulators essential for embryo development. *The Plant Cell Online, 15* (1 ), 5–18. Retrieved from http://www.plantcell.org/content/15/1/5.abstract

[67] Yamamoto, A., Kagaya, Y., Toyoshima, R., Kagaya, M., Takeda, S., Hattori, T. (2009). Arabidopsis NF-YB subunits LEC1 and LEC1-LIKE activate transcription by interacting with seed-specific ABRE-binding factors. *The Plant Journal: for Cell and Molecular Biology, 58* (5), 843–56. doi:10.1111/j.1365-313X.2009.03817.x

[68] Liu, W.-X., Zhang, F.-C., Zhang, W.-Z., Song, L.-F., Wu, W.-H., Chen, Y.-F. (2013). Arabidopsis Di19 functions as a transcription factor and modulates PR1, PR2, and PR5 expression in response to drought stress. *Molecular Plant, 6* (5), 1487–502. doi:10.1093/mp/sst031

[69] Song, S., Chen, Y., Zhao, M., Zhang, W.-H. (2012). A novel *Medicago truncatula* HD-Zip gene, MtHB2, is involved in abiotic stress responses. *Environmental and Experimental Botany, 80*, 1–9. doi:10.1016/j.envexpbot.2012.02.001

[70] Nakashima, K., Takasaki, H., Mizoi, J., Shinozaki, K., Yamaguchi-Shinozaki, K. (2012). NAC transcription factors in plant abiotic stress responses. *Biochimica et Biophysica Acta, 1819* (2), 97–103. doi:10.1016/ j.bbagrm.2011.10.005

[71] Nakaminami, K., Matsui, A., Shinozaki, K., Seki, M. (2012). RNA regulation in plant abiotic stress responses. *Biochimica et biophysica acta, 1819*(2), 149–53. doi:10.1016/j.bbagrm.2011.07.015

[72] Guleria, P., Mahajan, M., Bhardwaj, J., Yadav, S. K. (2011). Plant small RNAs: biogenesis, mode of action and their roles in abiotic stresses. *Genomics, Proteomics & Bioinformatics, 9* (6), 183–99. doi:10.1016/ S1672-0229(11)60022-3

[73] Yang, T., Xue, L., An, L. (2007). Functional diversity of miRNA in plants. *Plant Science, 172*(3), 423–432. doi:10.1016/j.plantsci.2006.10.009

[74] Lu, X.-Y., Huang, X.-L. (2008). Plant miRNAs and abiotic stress responses. *Biochemical and Biophysical Research Communications, 368*(3), 458–62. doi:10.1016/j.bbrc.2008.02.007

[75] Sunkar, R., Zhu, J. (2004). Novel and stress-regulated microRNAs and other small RNAs from Arabidopsis, *The Plant Cell, 16*, 2001–2019. doi:10.1105/tpc.104.022830.The

[76] Bartel, D. P., Lee, R., Feinbaum, R. (2004). MicroRNAs : Genomics , Biogenesis , Mechanism , and Function Genomics : The miRNA Genes, *Cell, 116*, 281–297.

[77] Sunkar, R., Girke, T., Jain, P. K., Zhu, J.-K. (2005). Cloning and characterization of MicroRNAs from rice. *The Plant Cell Online , 17* (5), 1397–1411.Retrieved from http://www.plantcell.org/content/17/5/1397.abstract

[78] Ding, D., Zhang, L., Wang, H., Liu, Z., Zhang, Z., Zheng, Y. (2009). Differential expression of miRNAs in response to salt stress in maize roots. *Annals of botany, 103*(1), 29–38. doi:10.1093/aob/mcn205

[79] Billoud, B., De Paepe, R., Baulcombe, D., Boccara, M. (2005). Identification of new small non-coding RNAs from tobacco and Arabidopsis. *Biochimie, 87*(9-10), 905–910. doi:10.1016/j.biochi.2005.06.001

[80] Arenas-Huertero, C., Pérez, B., Rabanal, F., Blanco-Melo, D., De la Rosa, C., Estrada-Navarrete, G., Sanchez, F., Covarrubias, A. A., Reyes, J. L. (2009). Conserved and novel miRNAs in the legume Phaseolus vulgaris in response to stress. *Plant Molecular Biology, 70* (4), 385–401. doi:10.1007/s11103-009-9480-3

[81] Qiu, C. X., Xie, F. L., Zhu, Y. Y., Guo, K., Huang, S. Q., Nie, L., Yang, Z. M. (2007). Computational identification of microRNAs and their targets in *Gossypium hirsutum* expressed sequence tags. *Gene, 395*(1-2), 49–61. doi:10.1016/j.gene.2007.01.034

[82] Xie, F. L., Huang, S. Q., Guo, K., Xiang, A. L., Zhu, Y. Y., Nie, L., Yang, Z. M. (2007). Computational identification of novel microRNAs and targets in Brassica napus. *Federation of European Biochemical Societies. letters, 581*(7), 1464–74. doi:10.1016/j.febslet.2007.02.074

[83] Yao, Y., Ni, Z., Peng, H., Sun, F., Xin, M., Sunkar, R., Zhu, J-K., Sun, Q. (2010). Non-coding small RNAs responsive to abiotic stress in wheat (*Triticum aestivum* L.). *Functional & Integrative Genomics, 10*(2), 187–190. doi:10.1007/s10142-010-0163-6

[84] Zhao, B., Liang, R., Ge, L., Li, W., Xiao, H., Lin, H., Ruan, K., Jin, Y. (2007). Identification of drought-induced microRNAs in rice. *Biochemical and Biophysical Research Communications, 354* (2), 585–90. doi:10.1016/j.bbrc.2007.01.022

[85] Lu, S., Sun, Y.-H., Shi, R., Clark, C., Li, L., Chiang, V. L. (2005). Novel and mechanical stress–responsive MicroRNAs in *Populus trichocarpa* that are absent from Arabidopsis. *The Plant Cell Online , 17* (8 ), 2186–

2203. Retrieved from http://www.plantcell.org/content/17/8/2186.abstract

[86] Ferguson, D. L., Guikema, J., Paulsen, G. M. (1990). Ubiquitin pool modulation and protein degradation in wheat roots during high temperature stress. *Plant Physiology*, *92* (3), 740–6. Retrieved from http://www.pubmedcentral.nih.gov/articlerender.fcgi?artid=1062362&tool=pmcentrez&rendertype=abstract

[87] Ortiz, C., Cardemil, L. (2001). Heat-shock responses in two leguminous plants: a comparative study. *Journal of Experimental Botany*, *52*(361), 1711–9. Retrieved from http://www.ncbi.nlm.nih.gov/pubmed/11479337

[88] Lyzenga, W. J., Stone, S. L. (2012). Abiotic stress tolerance mediated by protein ubiquitination. *Journal of Experimental Botany*, *63* (2), 599–616. doi:10.1093/jxb/err310

[89] Bartels, D., Sunkar, R. (2005). Drought and Salt Tolerance in Plants. *Critical Reviews in Plant Sciences*, *24 (1)*, 23–58. doi:10.1080/07352680590910410

[90] Foyer, C. H., Noctor, G. (2005). Redox homeostasis and antioxidant signaling: A metabolic interface between stress perception and physiological Responses. *The Plant Cell Online*, *17* (7), 1866–1875. Retrieved from http://www.plantcell.org/content/17/7/1866.short

[91] Mittler, R. (2002). Oxidative stress, antioxidants and stress tolerance. *Trends in Plant Science*, *7* (9), 405-410. http://linkinghub.elsevier.com/retrieve/pii/S1360138502023129

[92] Pitcher, L. H., Zilinskas, B. a. (1996). Overexpression of copper/zinc superoxide dismutase in the cytosol of transgenic Tobacco confers partial resistance to ozone-induced foliar necrosis. *Plant Physiology*, *110* (2), 583–588. http://www.pubmedcentral.nih.gov/articlerender.fcgi?artid=157754&tool=pmcentrez&rendertype=abstract

[93] Gupta, A. Sen, Webb, R. P., Holaday, A. S., Allen, R. D., Sciences, B. (1993). Overexpression of superoxide dismutase protects plants from oxidative Stress, *Plant Physiology, 103*, 1067–1073.

[94] Sunkar, R., Kapoor, A., Zhu, J. (2006). Posttranscriptional induction of two Cu / Zn Superoxide dismutase genes in Arabidopsis is mediated by downregulation of miR398 and important for oxidative stress tolerance, *The Plant Cell, 18*, 2051–2065. doi:10.1105/tpc.106.041673.1

[95] Borsani, O., Zhu, J., Verslues, P. E., Sunkar, R., Zhu, J.-K. (2005). Endogenous siRNAs derived from a pair of natural cis-antisense transcripts regulate salt tolerance in Arabidopsis. *Cell, 123* (7), 1279–91. doi:10.1016/j.cell.2005.11.035

[96] Zhu, J. K., Hu, X., Zhu, J-H. (2007). Role of microRNA in plant salt tolerance. United States Patent 20070214521.

[97] Zhao, B., Ge, L., Liang, R., Li, W., Ruan, K., Lin, H., Jin, Y. (2009). Members of miR-169 family are induced by high salinity and transiently inhibit the NF-YA transcription factor. *BMC Molecular Biology, 10,* 29. doi:10.1186/1471-2199-10-29

[98] Jung, H. J., Kang, H. (2007). Expression and functional analyses of microRNA417 in Arabidopsis thaliana under stress conditions. *Plant physiology and biochemistry: PPB / Société française de physiologie végétale, 45* (10-11), 805–11. doi:10.1016/j.plaphy.2007.07.015

[99] Vierstra, R. D. (2003). The ubiquitin/26S proteasome pathway, the complex last chapter in the life of many plant proteins. *Trends inPplant Science, 8* (3), 135-142 http://linkinghub.elsevier.com/retrieve/pii/S1360138503000141

[100] Kepinski, S., Leyser, O. (2005). The Arabidopsis F-box protein TIR1 is an auxin receptor. *Nature, 435* (7041), 446–451. Retrieved from http://dx.doi.org/10.1038/nature03542

[101] Reyes, J. L., Chua, N.-H. (2007). ABA induction of miR159 controls transcript levels of two MYB factors during Arabidopsis seed germination. *The Plant Journal: for Cell and Molecular Biology, 49*(4), 592–606. doi:10.1111/j.1365-313X.2006.02980.x

[102] Khraiwesh, B., Zhu, J.-K., Zhu, J. (2012). Role of miRNAs and siRNAs in biotic and abiotic stress responses of plants. *Biochimica et Biophysica Acta, 1819* (2), 137–48. doi:10.1016/j.bbagrm.2011.05.001

[103] Barrera-Figueroa, B. E., Wu, Z., Liu, R. (2012). Abiotic stress-associated microRNAs in plants: discovery, expression analysis, and evolution. *Frontiers in Biology, 8* (2), 189–197. doi:10.1007/s11515-012-1210-6

[104] Devi, S. J. S. R., Madhav, M. S., Kumar, G. R., Goel, a K., Umakanth, B., Jahnavi, B., & Viraktamath, B. C. (2013). Identification of abiotic stress miRNA transcription factor binding motifs (TFBMs) in rice. *Gene, 531* (1), 15–22. doi:10.1016/j.gene.2013.08.060

[105] Urano, K., Kurihara, Y., Seki, M., Shinozaki, K. (2010). "Omics" analyses of regulatory networks in plant abiotic stress responses. *Current Opinion in Plant Biology, 13* (2), 132–8. doi:10.1016/j.pbi.2009.12.006

[106] Matsui, A., Nguyen, A. H., Nakaminami, K., Seki, M. (2013). Arabidopsis non-coding RNA regulation in abiotic stress responses. *International Journal of Molecular Sciences, 14* (11), 22642–54. doi:10.3390/ijms141122642

[107] Matsui, A., Ishida, J., Morosawa, T., Mochizuki, Y., Kaminuma, E., Endo, T. A., Okamoto, M., Nambara, E., Nakajima, M., Kawashima, M., Satou, M., Kim, J-M., Kobayashi, N., Toyoda, T., Shinozaki, K., Seki, M. (2008). Arabidopsis transcriptome analysis under drought, cold, high-salinity and ABA treatment conditions using a tiling array. *Plant and Cell Physiology*, *49* (8),1135–1149. Retrieved from http://pcp.oxfordjournals.org/content/49/8/1135.abstract

[108] Urano, K., Kurihara, Y., Seki, M., Shinozaki, K. (2010). "Omics" analyses of regulatory networks in plant abiotic stress responses. *Current Opinion in Plant Biology*, *13* (2), 132–138. doi:10.1016/j.pbi.2009.12.006

[109] Sokol, A., Kwiatkowska, A., Jerzmanowski, A., Prymakowska-Bosak, M. (2007). Up-regulation of stress-inducible genes in tobacco and Arabidopsis cells in response to abiotic stresses and ABA treatment correlates with dynamic changes in histone H3 and H4 modifications. *Planta*, *227* (1), 245–54. doi:10.1007/s00425-007-0612-1

[110] Mirouze, M., Paszkowski, J. (2011). Epigenetic contribution to stress adaptation in plants. *Current Opinion in Plant Biology*, *14* (3), 267–274. doi:10.1016/j.pbi.2011.03.004

[111] Reik, W., Dean, W., Walter, J. (2001). Epigenetic reprogramming in mammalian development. *Science*, *293* (5532), 1089–1093. doi:10.1126/science.1063443

[112] Tsukahara, S., Kobayashi, A., Kawabe, A., Mathieu, O., Miura, A., Kakutani, T. (2009). Bursts of retrotransposition reproduced in Arabidopsis. *Nature*, *461* (7262), 423–6. doi:10.1038/nature08351

[113] Mirouze, M., Reinders, J., Bucher, E., Nishimura, T., Schneeberger, K., Ossowski, S., Cao, J., Weigel, D., Paszkowski J., Mathieu, O. (2009). Selective epigenetic control of retrotransposition in Arabidopsis. *Nature*, *461* (7262), 427–30. doi:10.1038/nature08328

[114] Paszkowski, J., Whitham, S. A. (2001). Gene silencing and DNA methylation processes. *Current Opinion in Plant Biology*, *4* (2), 123–9. Retrieved from http://www.ncbi.nlm.nih.gov/pubmed/11228434

[115] Akimoto, K., Katakami, H., Kim, H.-J., Ogawa, E., Sano, C. M., Wada, Y., Sano, H. (2007). Epigenetic inheritance in rice plants. *Annals of Botany*, *100* (2), 205–17. doi:10.1093/aob/mcm110

[116] Steward, N., Ito, M., Yamaguchi, Y., Koizumi, N., Sano, H. (2002). Periodic DNA methylation in maize nucleosomes and demethylation by environmental stress. *The Journal of Biological Chemistry*, *277* (40), 37741–6. doi:10.1074/jbc.M204050200

[117] Boyko, A., Kovalchuk, I. (2011). Genome instability and epigenetic modification-heritable responses to environmental stress? *Current Opinion in Plant Biology, 14* (3), 260–6. doi:10.1016/j.pbi.2011.03.003

[118] Molinier, J., Ries, G., Zipfel, C., & Hohn, B. (2006). Transgeneration memory of stress in plants. *Nature, 442* (7106), 1046–9. doi:10.1038/nature05022

[119] DeBolt, S. (2010). Copy number variation shapes genome diversity in Arabidopsis over immediate family generational scales. *Genome Biology and Evolution, 2,* 441–53. doi:10.1093/gbe/evq033

[120] Maccaferri, M., Sanguineti, M. C., Corneti, S., Ortega, J. L. A., Salem, M. Ben, Bort, J., DeAmbrogio, E., Garcia del Moral, L.F., Demontis, A., El-Ahmed, A., Maalouf, F., Machlab, H., Martos,V., Moragues, M., Motawaj, J., Nachit, M., Nserallah, N., Ouabbou, H., Royo, C., Slama, A., Tuberosa, R. (2008). Quantitative trait loci for grain yield and adaptation of durum wheat (*Triticum durum* Desf.) across a wide range of water availability. *Genetics, 178* (1), 489–511. doi:10.1534/genetics.107.077297

[121] Cooper, M., van Eeuwijk, F. A, Hammer, G. L., Podlich, D. W., Messina, C. (2009). Modeling QTL for complex traits: detection and context for plant breeding. *Current Opinion in Plant Biology, 12* (2), 231–40. doi:10.1016/j.pbi.2009.01.006

[122] Montes, J. M., Melchinger, A. E., Reif, J. C. (2007). Novel throughput phenotyping platforms in plant genetic studies, *TRENDS in Plant Science 12* (10), 10–13.

[123] Parent, B., Turc, O., Gibon, Y., Stitt, M., Tardieu, F. (2010). Modelling temperature-compensated physiological rates, based on the co-ordination of responses to temperature of developmental processes. *Journal of Experimental Botany, 61* (8), 2057–69. doi:10.1093/jxb/erq003

[124] Chenu, K., Chapman, S. C., Tardieu, F., McLean, G., Welcker, C., Hammer, G. L. (2009). Simulating the yield impacts of organ-level quantitative trait loci associated with drought response in maize: a "gene-to-phenotype" modeling approach. *Genetics, 183* (4), 1507–23. doi:10.1534/genetics.109.105429

[125] Collins, N. C., Tardieu, F., Tuberosa, R. (2008). Quantitative trait loci and crop performance under abiotic stress: where do we stand? *Plant physiology, 147* (2), 469–86. doi:10.1104/pp.108.118117

[126] Venuprasad, R., Dalid, C. O., Del Valle, M., Zhao, D., Espiritu, M., Sta Cruz, M. T., Amante, M., Kumar, A., Atlin, G. N. (2009). Identification and characterization of large-effect quantitative trait loci for grain yield

under lowland drought stress in rice using bulk-segregant analysis. *TAG. Theoretical and applied genetics. Theoretische und angewandte Genetik, 120*(1), 177–90. doi:10.1007/s00122-009-1168-1

[127] Ma, H-S, Liang, D., Shuai, P., Xia, X-L., Yin, W-L.. (2010). The salt- and drought-inducible poplar GRAS protein SCL7 confers salt and drought tolerance in *Arabidopsis thaliana. Journal of Experimental Botany, 61*, 4011–4019.

[128] Qin, Y., Wang, M., Tian, Y., He, W., Han, L., Xia, G. (2012) Over-expression of TaMYB33 encoding a novel wheat MYB transcription factor increases salt and drought tolerance in Arabidopsis. *Molecular Biology Reports, 39 (6)*, 7183-7192.

[129] Prabu, G., Prasad, D. T., (2012). Functional characterization of sugarcane MYB transcription factor gene promoter (PScMYBAS1) in response to abiotic stresses and hormones. *Plant Cell Reports, 31*, 661–669.

[130] Shin, D., Moon, S-J., Han, S., Kim, B-G., Park, S.R., Lee, S-K., Yoon, H-J., Lee, H. E., Kwon, H-B., Bu, D. B., Yi, Y., Byun, M. O.(2010). Expression of StMYB1R-1, a novel potato single MYB-like domain transcription factor, increases drought tolerance. *Plant Physiology, 155*, 421–432.

[131] Mare, C., Mazzucotelli, E., Crosatti, C., Francia, E., Stanca, AM., Cattivelli, L. (2004) Hv-WRKY38: a new transcription factor involved in cold- and drought-response in barley. *Plant Molecular Biology, 55*, 399–416.

[132] Kim, J. C., Lee, S. H., Cheong, Y. H., Yoo, C. M., Lee, S. I., Chun, H. J., Yun, D. J., Lee. S. Y., Lim, C. O., Cho, M. J. (2001). A novel cold-inducible zinc finger protein from soybean, SCOF-1, enhances cold tolerance in transgenic plants. *Plant Journal, 25,* 247–259.

[133] Sakamoto, H., Araki, T., Meshi, T., and Iwabuchi, M. (2000). Expression of a subset of the Arabidopsis Cys(2)/His(2)-type zinc-finger protein gene family under water stress. *Gene, 248*, 23–32.

[134] Xu, S., Wang, X., and Chen, J. (2007). Zinc finger protein 1 (ThZF1) from salt cress (*Thellungiella halophila*) is a Cys-2/His-2-type transcription factor involved in drought and salt stress. *Plant Cell Reports, 26*, 497–506.

[135] Harris, J. C., Hrmova, M., Lopato, S., Langridge, P. (2011) Modulation of plant growth by HD-Zip class I and II transcription factors in response to environmental stimuli. *New Phytologist, 190(4)* 823-837.

In: Abiotic Stress                    ISBN: 978-1-63117-622-7
Editor: Annabella Ferro              © 2014 Nova Science Publishers, Inc.



# Chapter 2

# STRUCTURAL ASPECTS AND FUNCTIONAL REGULATION OF LATE EMBRYOGENESIS ABUNDANT (LEA) GENES AND PROTEINS CONFERRING ABIOTIC STRESS TOLERANCE IN PLANTS

*Aryadeep Roychoudhury*[*] *and Subhayu Nayek*
Post Graduate Department of Biotechnology, St. Xavier's College
(Autonomous), 30, Mother Teresa Sarani, Kolkata, West Bengal, India

## ABSTRACT

Late embryogenesis abundant (LEA) proteins are the members of a large group of hydrophilic, low molecular weight (10–30 kDa) proteins found primarily in plants, encoded by multigene families. They were first characterized in cotton and wheat and are produced in abundance, late during embryo development, constituting around 4% of the total cellular proteins. They are mostly localized in cytoplasm and nuclear region. Genes of LEA proteins have been identified in many plant species; their expression is linked to the acquisition of desiccation tolerance in orthodox seeds, pollen and anhydrobiotic plants. However, many LEA proteins are induced by salinity, drought, cold or other osmotic stress and by exogenous abscisic acid (ABA) in vegetative tissues as well, where

---

[*] E-mail: aryadeep.rc@gmail.com, aryadeep19@rediffmail.com.

they play a major role as cellular protectants. These proteins are part of evolutionarily conserved group of hydrophilic proteins termed "hydrophilins" involved in various adaptive responses to hyperosmotic conditions. Although their precise function is still obscure, a number of putative mechanisms have been proposed to LEA proteins where they act as either hydrating buffer, molecular shield by sequestering ions, as chemical chaperones by helping in renaturing unfolded cellular proteins, and transport of nuclear targeted proteins during stress. The majority of LEA proteins display preponderance of hydrophilic and charged amino acid residues, lacking or having a low proportion of Cys and Trp residues. At least six different groups of LEA proteins have been identified based on their amino acid sequence, mRNA homology and expression pattern; the major categories being group 1, group 2, group 3, group 4 and group 5. LEA protein synthesis, expression and biological activities are regulated by many factors (e.g., developmental stages, hormones, ion change and dehydration) and signal transduction pathways. The best characterized cis-element in *LEA* genes with context to ABA-mediated osmotic stress induction is the ABA-responsive element (ABRE), which contains the palindromic motif CACGTC. The ABREs can interact mostly with the basic leucine zipper (bZIP) transcription factors. A 9-bp conserved sequence, TACGACAT, termed dehydration responsive element (DRE) or C-repeat (CRT) elements has been reported in the promoter regions of ABA-independent cold- and drought-inducible genes. Transgenic plants overexpressing several *LEA* genes under constitutive or abiotic stress-inducible promoters have been reported to show enhanced tolerance to single or multiple stress conditions. Since the study of the regulatory mechanism of *LEA* gene expression is an important feature of stress molecular biology, a detailed discussion on structure, function and expression regulation of LEA proteins in higher plants is presented in this chapter.

# 1. INTRODUCTION

The higher plants have developed multi-pathway, multi-level and multi-scale survival strategies for continual changes in the environment, which include anatomical, physiological, biophysical, biochemical, genetic and developmental changes in response to adverse conditions. Abiotic stress such as drought, salinity, high or low temperature and heavy metals cause major damage and decreased yield of plants. It has been estimated that 70% of the crop yield loss can be attributed to abiotic stresses, especially drought (Bray et al. 2000). Under severe condition, these adverse environmental stresses can result in death of plant. Plants must respond and adapt to these adverse conditions to avoid or decrease cell injury. During cell drying, the accumulation of macromolecules such as oligosaccharides or proteins greatly increases cytoplasmic viscosity and usually causes the formation of bioglasses. Thus, bioglasses have been suggested to provide intracellular protection against the denaturation of large molecules to stabilize plasma membranes. The major seed proteins involved in bioglass formation are the late embryogenesis abundant (LEA) proteins (Buitink and Leprince 2004). The evolution of LEA proteins and their functional expression is thus one of the adaptive changes, which plays an important role in resistance to various environmental stresses (Hong-Bo et al. 2005). This implies that studies on LEA proteins, and the isolation, identification and functional analysis of their genes will be of great benefit to the generation of abiotic stress-resistant crops, a major challenge for crop scientists. These efforts have been greatly aided by the sequencing of the whole genomes of *Arabidopsis thaliana* and *Oryza sativa*. Here, we discuss the role of LEA protein structure, function and gene expression in the development of abiotic stress-resistant crops.

## 2. DISTRIBUTION LEA PROTEINS IN HIGHER PLANTS

LEA proteins are found in both angiosperms and gymnosperms, as well as in moss and pteridophytes (Battaglia and Covarrubias 2013). LEA proteins are also found in a wide range of other organisms including bacteria (Stacy et al. 1999), yeast (Garay-Arroyo et al. 2000), cyanobacteria (Tanaka et al. 2004), nematode (Solomon et al. 2000), brine shrimp (Sharon et al. 2009) and collembolan (Bahrndorff et al. 2009). In plants, the LEA proteins were first isolated in developing cotton seeds. Subsequently, researchers detected their existence in different plant species like wheat, barley, maize, rice, sunflower, potato, grape, apple, bean, *Arabidopsis*, tomato, rye, soybean, carrot and so on. As the name suggests, LEA proteins are late embryonic proteins abundant in higher plant embryos and accumulate to high levels during the last stage of seed maturation or late period of seed development, accompanied by dehydration, but disappear following germination (Dure 1993). They have also been noted in seedlings, roots and other vegetative tissues after induction with exogenous abiotic stresses, suggesting a protective role (Ingram and Bartels 1996). Some LEA proteins are associated with vascular tissues and meristematic regions (Moreno-Fonseca and Covarrubias 2001). LEA proteins are part of evolutionarily conserved group of hydrophilic proteins termed "hydrophilins" involved in various adaptive responses to hyperosmotic conditions (Garay-Arroyo et al. 2000). They are proteins with small molecular weight ranging mainly from 10 to 30 kDa, localised in the cytoplasm, nucleus, mitochondria, vacuoles, near the cellular membrane, protein bodies (Grelet et al. 2005) as well as in amyloplasts (Rinne et al. 1999). Thus, they appear to be located in many cell types and at variable concentrations, and within the cell, they appear to be predominantly but not exclusively cytosolic. After synthesis, they are transported to cell organelles and membranes, where they stabilize cell structures and molecules. The partition of LEA proteins into various cell compartments determines their possible function in the protection or regulation of essential biochemical processes, like replication or respiration. Their concentrations in the cell are characteristically very high. For example, in mature cotton embryo cells, the D7 LEA proteins represent about 4% of non-organellar cytosolic protein (about 0.34 mM) (Ingram and Bartels 1996). The scrutiny of soybean microarray information showed that *LEA* transcripts from different groups accumulate in root tips and apical meristems. While soybean *LEA6* transcripts were not detected in meristems, the transcript and protein for its homologous gene accumulate in vegetative and root meristems from principal and lateral roots of *Phaseolus vulgaris* (Moreno-Fonseca and Covarrubias 2001). These observations strongly support the idea that LEA proteins could protect stem cell niches from changes in environmental water availability preventing and/or avoiding damage to these cells which is vital for plant survival.

## 3. STRUCTURAL AND FUNCTIONAL BASIS OF CLASSIFICATION OF LEA PROTEINS IN PLANTS

The physical characteristics of LEA proteins may provide some clues to their mode of action. Several nomenclature systems have been reported to classify LEA proteins into different groups. A more common classification of *LEA* genes was then projected by deduced protein structural domains or chemical characteristics. The traditional criterion was initially described by Dure et al. (1989). Groupings for dividing the LEA proteins originated from a dot matrix analysis with proteins from cotton. In early studies, LEA proteins were classified according to their molecular weight and therefore were named as D7, D11, D19, D29, D34, D73, D95, D113 (Roychoudhury and Paul 2012). A group was assigned on the basis of one cotton LEA protein showing regions of significant homology with at least one protein from another species. The "type" of cotton proteins used for these groupings were LEA D19 (Group 1), LEA D11 [Group 2 (also termed dehydrins)], and LEA D7 (Group 3). The cotton proteins LEA D113 and LEA D95 define two additional classes. As more other LEA proteins were identified, new classification or nomenclature systems based on protein amino acid composition or different sequence motifs was applied, on the basis of which LEA proteins were classified into five to nine subclasses (Bies-Etheve et al. 2008). This system will remain useful until clear functions can be assigned. Another criterion was provided by a computer analysis, "Protein or Oligonucleotide Probability Profile" (POPP), which showed over- or under- representation of particular amino acids in protein sequences (Wise 2003). After clustering by consensus POPP, LEA proteins were reclassified into at least four groups. Despite the different criteria of classification, the primary structures of most LEA proteins shared similar biophysical features. The Kyte–Doolittle hydropathy algorithm revealed that most LEA proteins were hydrophilic (Baker et al. 1988).

A general structural feature of the LEA proteins is their biased amino acid composition, so that the majority of LEA proteins display preponderance of hydrophilic and charged amino acid residues, ordered in repeated sequence (e.g., Gly and Lys), forming hyper-hydrophilicness and thermal stability; for example, a deduced D19 protein from cotton contains 13% Gly and 11% Glu. Furthermore, most LEA proteins have a low proportion of Cys, Trp and Val residues. In general, they are randomly coiled proteins in solution and therefore considered as intrinsically unstructured proteins (Battaglia et al. 2008). Advanced structure of such protein contains non-periodic linear and α-helix structure without thermal dominative state and corresponding dehydrated proteins exist in a natural form of dimers. There are a few atypical LEA proteins that contain a significantly higher proportion of hydrophobic residues and predicted to adopt more globular conformations. For examples, the cotton protein D34, D73, D95 and a D95-like protein from *Arabidopsis* all belong to the atypical LEA proteins.

At least six groups of LEA proteins have been categorized by virtue of similarities in their deduced amino acid sequences and the first three groups are the major ones. More common classifications are related to their protein structural domains or chemical characteristics like the appearance of different sequence motifs/patterns or biased amino acid composition.

## Group 1

Group 1 proteins (e.g., *Gossypium hirsutum* D19, *Triticum aestivum* Em, *Hordeum vulgare* B19 and sunflower HaDs10) are characterized by an internal 20-amino-acid conserved signature motif [(R/G)S(R/K) GGQTRKEQLGXEGTXEM] near the C-terminus, repeated up to four times depending on the species and a high proportion of Gly, Glu, Gln and charged amino acid residues, e.g., *Arabidopsis* has two *Em* genes, *AtEm1*, which contain four repeats of 20-mer motifs, and *AtEm6*, with one 20-mer motif. They can absorb a large amount of bound water and hence help in water binding to provide a protective aqueous environment for the cellular components. The conformation is predominantly random coil with some predicted short α-helices. They contain more water of hydration than typical globular proteins. Group 1 protein genes play a role in endosperm development and osmotic protection of vegetative organs in higher plants (Hong-Bo et al. 2005).

## Group 2

Group 2 proteins (e.g., *Gossypium hirsutum* D11, maize DHN-1, M3 and Rab17, *Arabidopsis* pRABAT1, ERD10 and ERD14, *Craterostigma* pcC27-04 and pcC6-19, tomato pLE4 and TAS14, barley B8, B9 and B17 and carrot pcEP40) also referred to as dehydrins, are characterized by a highly conserved 15-amino-acid Gly and Lys-rich sequence in the C-terminus with molecular weight of 9.2 kDa (Close 1997). Gly, His, Lys and Thr comprise 56% of all amino acids in LTI30 (water stress-responsive protein). The K, S and Y segments are characteristic of dehydrins (Close 1996; Brini et al. 2011). The K-segment, rich in Lys (K) residues, has a consensus repetitive 15-amino acid sequence EKKGIMDKIKEKLPG at the C-terminus. Substitutes and structural modifications of amino acids are possible in the K segment, which can be repeated 1.12 times in a single protein. The K-segment can form an amphipathic hydrophilic α-helix structure which is probably involved in protein-protein and protein-lipid interaction. Experimental evidence showed that K-segment was important to stabilize other cellular components under stress condition (Koag et al. 2009). Many dehydrins contain a stretch of contiguous Ser or Thr residues called S segment, which act as the phosphorylation site of the protein. The S segment phosphorylation by casein kinase II (CKII) or other Ser/Thr protein kinases might be an important posttranslational modification, which is required for proper protein localization in the nucleus and nucleolus, as it is observed during Rab17 placing in the nucleus (Goday et al. 1994). The S-segment and RRKK sequence were considered to be the nuclear localization signal (Jensen et al. 1998). When present, the S-segment usually precedes the K-segment (Campbell and Close 1997). Moreover, these motifs were embedded in amorphous regions with Gly/Ser-, Gly/Thr-, or Glu/Lys- rich residues designated as $\varphi$ domains. However, not every S segment containing dehydrin has the function of phosphorylation-dependent $Ca^{2+}$ binding. The neutral LEA2 protein DHR18 contains the S segment but has no $Ca^{2+}$ binding activity (Alsheikh et al. 2005). The Y fragment, located at the N-terminus, has the consensus sequence of (V/T) DEYGNP, and shows similarity to nucleic acid binding sites of chaperone proteins from bacteria and plants (Close 1997). In addition to these motifs previously identified, a new Gly-rich region was found in some legume LEA2 proteins. This region showed a repetitive large motif represented by the sequence [YF]T[GD]DT[GN] [RK]QH[GD]T and is present up to ten times in a protein, as in the case of the polypeptide encoded by the *Phvul004G158800.1* gene from *Phaseolus vulgaris*. Two dehydrins with Gly-rich motifs have been described. One from *Vigna radiata*, VrDhn1, a novel $Y_2K$ dehydrin detected in seeds and induced by various abiotic stresses or ABA treatment (Lin et al. 2012), and Mat1 (Glyma07g10030.1) from soybean, which has a homologue called Mat9 (Glyma09g31740.2), induced upon severe water deficit (Whitsitt et al. 1997; Momma et al. 2003). This motif was only detected in LEA2 proteins from *Phaseoloid* species, which include several legumes more adapted to tropical climates. This group is one of the youngest groups among the three legume sub-families (Gepts et al. 2005). So it can be suggested that this motif is of recent appearance. It is worth mentioning that this motif and its repetitions (4–10X) were always found to be present between segments-Y and-K, which suggests a common ancestor for this type of dehydrins.

The classification of dehydrins is based on the presence of conserved segments. Five classes of dehydrins can be distinguished: $Y_nSK_2$, $K_n$, $K_nS$, $SK_n$ and $Y_2K_n$, where n is the number of repeat. The wide range of molecular weight in LEA2 proteins is caused by the repeat number of K segments. The $Y_nSK_2$ dehydrins consist of 1.3 Y segments, a Ser-rich segment, and two K segments. Their expression is induced mainly by ABA (Rab type) and drought, less often by low temperature. Some $Y_nSK_2$ dehydrins have an ability to bind lipid vesicles containing acidic phospholipids (Koag et al. 2003). In $K_n$ dehydrins, the occurrence of 1.9 Ser-rich segments is observed. They are acidic or neutral proteins, mainly induced by low temperature and somewhat by ABA and drought. The K segment of class $K_nS$ begins with sequence (H/Q) KEG instead of EKKG. Their expression is induced by ABA and dehydration. The $K_nS$ dehydrins can bind various metal ions or scavenge the hydroxyl radicals (Ashgar et al. 1994). $Y_2K_n$ proteins usually consist of two Y and two K segments. They are expressed during seed germination and related with resistance to low temperature. The $SK_n$ dehydrins contain one S and 1.3 K segments and are preferentially induced by low temperature. Moreover, they can be induced by drought, salinity, injury and jasmonate treatment. Dehydrins lacking the Y segment were identified in many other plant tissues, such as stem tissue, vegetative and reproductive buds, needle tissue, unstressed leaf and root tissue. Its expression was induced by water, salt, cold and osmotic stress (Caruso et al. 2002). The K segment was considered to appear in all classes of dehydrins, but recently a protein lacking the K segment (ROD25 from cold acclimated *Cornus sericea*, syn. *C. stolonifera*) was detected (Sarnighausen et al. 2002). Hinniger et al. (2006) characterized several full-length cDNA-encoding dehydrins (CcDH1, CcDH2 and CcDH3) and a LEA protein (CcLEA1) from *Coffea canephora* (robusta). The *CcDH1* and *CcDH2* genes encode $Y_3SK_2$ dehydrins and the *CcDH3* gene encodes an $SK_3$ dehydrin. *CcDH1* and *CcDH2* were found to express during the final stages of arabica and robusta grain development, but only the *CcDH1* transcripts were clearly detected in other tissues such as pericarp, leaves and flowers. *CcDH3* transcripts were also found in developing arabica and robusta grain, in addition to being detected in pericarp, stem, leaves and flowers. The role of all these conserved segments is to exert their biological function upon recognition and interaction with specific biological targets. The combination and number of repeated segments is related with the possible protein functions displayed in seeds under stress conditions. Dehydrins are thus induced by dehydration-related stresses such as low temperature, drought and high salinity and response to wounding. They play an important role in metabolism as molecular chaperones and defending protein structure with a close connection to higher plant resistance to drought. They also exclude solutes from the surface of membranes and cytosolic proteins, preventing denaturation and maintaining the solvation of structural surfaces.

## Group 3

Group 3 LEA proteins (e.g., cotton D7, *Hordeum vulgare* HVA1 and PMA1949, *Brassica* pLEA76, soybean pmGM2, wheat pMA2005 and pMA1949, *Daucus carota* Dc8 and Dc3 and *Craterostigma* pcC3-06) have a tandemly repeated 11-mer amino acid fragment (TAQAAKEKAGE, KAKETKDAAAE, TKDYAGDAAQK or φφE/QXφKE/QKφXE/D/Q, where φ represents a hydrophobic amino acid) in 13-repeats that may form an amphiphilic α-helix structure and are involved in enriching ions during dehydration of higher plants (Tunnacliffe and Wise 2007; Yang et al. 2013). A large variation in molecular weight among LEA3 proteins is caused by the number of tandem repeats, which range from 5 to more than 30. The average molecular weight of the LEA3 proteins is 25.5 kDa (maximum 67.2, minimum 7.2). These 11 mer motifs have been proposed to be linked by ionic bridges and play important physiological roles. These proteins exist as dimers, the polar face of these dimerized helices are exposed and capable of binding ions via formation of salt bridges. So they act in ion sequestration. Lea76 of *Brassica napus* contains 13 repeats of a homologous amino acid motif (Harada et al. 1989), and HVA1 of barley contains nine repeats (Hong et al. 1988). Group 3 LEA proteins with similar repeating amino acid motifs have also been reported in cotton, carrot and rapeseed (Harada et al. 1989; Zhao et al. 2010).

## Group 4

Group 4 proteins [e.g., LEA14, cotton D113, sunflower Hads11, tomato LE25, soybean GmPm1 and GmPm16, *Arabidopsis* AtLEA4-1 (PAP260) and AtLEA4-5 (PAP51) and *Phaseolus vulgaris* PvLEA-18], are devoid of repeated consensus motif sequences like LEA1, LEA2 and LEA3 proteins. They contain several conserved small and charged amino acids at N-terminal end to form amphipathic α-helix structure. The C-terminus is less conserved, rich in Gly and nonpolar amino acids or amino acids containing hydroxyl groups forming an unstructured random coil of variable length. The above proteins can further form a subsidiary structure adaptive to conformational changes of other proteins and function in protecting membrane stability and integration during dehydration (Chaves et al. 2003). LEA4 proteins contain high proportions of small amino acid residues such as Gly or Ala and charged ones such as Asp, Gly, Lys, or Arg and are thus highly hydrophilic. Thus, these proteins can bind water molecules and also act as reverse chaperones, whereby they stabilize membrane surfaces and proteins by functioning as solvation film. Like LEA 1, 2 and 3 proteins, LEA 4 proteins are heat soluble, that is, they do not precipitate after boiling for 10 min.

## Group 5

Group 5 [e.g., cotton D34, D73 and D95, maize RAB28, carrot DcECP31, *Arabidopsis* AtECP31 and AtRAB28, *Craterostigma plantagineum* pcC27-45 (expressed in response to salt in callus tissues and in response to ABA and desiccation in both leaves and callus), tomato ER5, hot pepper (*Capsicum frutescens*) CaLEA6, *Medicago truncatula* MtPM25, and soybean GmPM22, GmPM24, GmPM25, and GmPM26] proteins also have an 11-mer amino acid repeat, in which each amino acid has the chemical properties similar to Group 3, but lacks a high degree of amino acid-residue specificity at each position. Most LEA5 proteins are acidic. The members of this group are distinctively different from the other four groups of LEA proteins by their containing a high proportion of hydrophophic residues and lacking the feature of boiling solubility. Hence, LEA 1, 2, 3 and 4 proteins are the so called hydrophilic proteins, whereas LEA 5 proteins are hydrophobic, with the N-terminal end forming amphipathic α-helix (Galau et al. 1993). It is predicted that Group 5 proteins play roles in seed maturation, dehydration and combining concentrated ions (Chen et al. 2002; Yu et al. 2002). Naot et al. (1995) isolated a gene from the salt-tolerant line of Shamuti orange, which encode a group 5 LEA-homologue in response to salt and water deficit.

## Group 6

Group 6 or atypical LEA proteins, like other groups of LEA proteins accumulate in seeds during late embryogenesis and disappear after the early hours of germination. They may also accumulate in other organs during various environmental stresses. This group does not contain any motif or protein characteristics defined in other LEA proteins; examples are barley HVA22 (Shen et al. 1993), riceWSI76 (Takahashi et al. 1994), *Arabidopsis* AtDI19, AtLEA9-1 (M10) and AtLEA9-2 (M17) (Raynal et al. 1999), wheat AWPM-19 and WCOR413 (Breton et al. 2003; Koike et al. 1997) and soybean GmPM4 (Hsing et al. 1998). The expression pattern suggests that deduced products of these genes are associated with the development of late embryogenesis. However, unlike typical hydrophilic LEA proteins, they may be carrier proteins, galactinol synthases, fiber proteins or plasma membrane proteins. Thus, the members of this group fit with the primary definition of LEA proteins but have diverse protein characteristics and functions.

## 4. STRUCTURE-FUNCTION CORRELATION OF LEA PROTEINS

The secondary structures of several LEA proteins have been characterized by circular dichroism (CD), nuclear magnetic resonance (NMR), or Fourier transform infra-red (FTIR) spectroscopy. The amino acid composition of LEA proteins reveals that hydrophilic LEA proteins (LEA 1–4) are rich in small and hydrophilic amino acid residues and poor in non polar ones. By contrast, LEA 5 proteins and a small heat shock protein Hsp22, which contain defined secondary structures, have more non polar residues. The protein-folding parameters also suggest that all listed hydrophilic LEA proteins fit with unfolded ones. For LEA 5 proteins, the protein folding parameters of M88322 and cotton D95 fit with folded proteins. Thus, these parameters may be used to find out most *LEA* genes in a genome-wide prediction analysis. The consensus POPP analysis of various LEA proteins provides direct evidence to define hydrophilic LEA proteins as natively unfolded proteins. Most folded proteins contain a hydrophobic core with the side chain stabilizing the folded conformation and charged or polar side chains on the surface where they interact with surrounding water molecules. By contrast, natively unfolded proteins perform mainly the water molecule–protein interaction in lieu of intramolecularly hydrophobic interaction. Hence, they fail to obtain defined structures. Moreover, unfolded LEA proteins may carry additional water molecules. Natively unfolded proteins may change their conformation by interacting with other molecules such as proteins, nucleic acids or metal ions. Although hydrophilic LEA proteins interact with SDS micelles and perform the protein-folding process, no evidence exists to show that hydrophilic LEA proteins could interact with sugars, metal ions or even phospholipids in solution. However, LEA proteins did perform conformational changes during dehydration because superficial groups of proteins no longer bound hydrogen with water and instead formed inter- and intra- molecular hydrogen bonds with one another. With dehydration, the biochemical features of LEA proteins change. They start to interact with various oligosaccharides, suggesting that the proteins act synergistically with the oligosaccharides in the formation of the glassy matrix (Shih et al. 2008).

The exact biological functions of most conserved motifs in LEA proteins are unknown. The function of only a small number of motifs could be proposed in plants. For example, the N-terminal motif of LEA1 proteins and N-terminal motif of LEA3 protein, and the motif1 of LEA4 (Shih et al. 2004) have the ability to form α-helix which may play a protective role on functional proteins and maintain normal physiological processes in plant cells under water deficiency.

## 5. Probable Functions of LEA Proteins

The precise functions of LEA proteins are still rather enigmatic. We await direct experimental evidence that LEA proteins can protect specific cellular structures or ameliorate the effects of drought stress. Recent studies have shown that LEA proteins can improve plant tolerance to salt, drought, and/or freezing stresses through various physiological pathways. Because they are highly hydrophilic, it appears unlikely that they occur in specific cellular structures. Also, their high concentrations in the cell and biased amino acid compositions suggest that they do not function as enzymes. In view of the apparent lack of well-ordered tertiary structure of LEA proteins preventing their use as catalysts, several mechanisms have been proposed to relate their structural features to the protection of cellular structures required by a dehydrated state, viz., water replacement or water binding, ion sequestering, macromolecules and membrane stabilization and chaperone-like activity (Close 1996, 1997; Cuming 1999). Many mechanisms have been proposed for LEA proteins, such as hydration buffer, molecular shield, enzyme- protectant and membrane interaction.

1) The randomly coiled moieties of some LEA proteins are consistent with a role in binding water. Total desiccation is probably lethal and therefore such proteins could help maintain the minimum cellular water requirement. A working model is that hydration of the hydrophilic region of dehydrins results in the formation of an envelope of ordered water, which especially in the presence of compatible solutes, operates through a preferential exclusion mechanism to either drive a partially unfolded protein back into a folded state or at least inhibit further denaturation. The Em protein

(D19-group) from wheat is considerably more hydrated than most globular polypeptides because it is over 70% random coil under normal physiological conditions. The random coil tails of the D113 proteins could also bind considerable amounts of water, although the long N-terminal helical domain would not share this property (Li et al. 1998; Chaves et al. 2003).

2) Since LEA proteins are highly hydrophilic, they may also act as hydration buffers and decrease the water-loss rate during stress conditions. LEA proteins can stabilize the membranes by inducing preferential hydration or replacing water at the interaction interface with other macromolecules, during desiccation conditions. Most evidence for membrane stabilization comes from *in vitro* experiments, where LEA2 proteins were found associated to anionic phospholipid vesicles (Koag et al. 2003, 2009). Other assays also suggested that dehydrin addition maintain the functional membrane structure under freezing temperatures (Ismail et al. 1999a, b; Kosová et al. 2007). These data are in agreement with reports that have detected binding of dehydrins to membranes or phospholipids. A LEA protein from maize, DHN1, was demonstrated to undergo conformational changes when binding specifically lipid vesicles, suggesting a role in membrane stabilization during stress (Koag et al. 2003). However, all these results have been obtained from *in vitro* experiments or with isolated membranes. In contrast, no conclusive information is available regarding their membrane association by intracellular localization experiments using antibodies or fluorescent translational fusions, which however, have contributed to establish dehydrin location in cytosol, nuclei, chloroplast, protein and lipid bodies, mitochondria, and plasmodesmata (Rorat 2006). The LEA proteins D11 and D113 could be involved in the "solvation" of cytosolic structures. The random coiling would permit their shape to conform to that of other structures and provide a cohesive layer with possibly greater stability than would be formed by sugars. Their hydroxylated groups would solvate the structural surfaces.

3) Because LEA proteins have many charged amino acid residues, they can play a role in sequestering ions during water loss, viz., LEA2 and LEA4 proteins, where His-containing motifs seem to be responsible for binding divalent cations (Hara et al. 2005). The soybean LEA4-5 proteins          (GmPM1:Glyma19g32920.1          and          GmPM9: Glyma03g30040.1), can bind $Fe^{+3}$, $Ni^{+2}$, $Cu^{+2}$ and $Zn^{+2}$ (Liu et al.

2011). The 11-amino-acid motif (T/AA/TQ/EA/TA/ TK/RQ/EDK/RA/TXED/Q) of LEA protein D29 (which is also present in D7 LEA proteins) could counteract the irreversibly damaging effects of increasing ionic strength in the cytosol during desiccation. Such problems could be mitigated by the formation of salt bridges with amino acid residues of highly charged proteins. The repeating elements most likely exist as amphiphilic helices (Zhang and Zhao 2003) which means that hydrophobic and hydrophilic amino acids are contained in particular sectors of the helix. The helices probably form intramolecular bundles, which would present a surface capable of binding both anions and cations. Further analyses of the D7-group of molecules have allowed precise structural predictions to be made. The intersurface edges of the interacting helical regions of the (putative) dimer reveal periodically spaced binding sites for suitably charged ions. Blackman et al. (1995) demonstrated that the increased LEA protein level might reduce the electrolyte leakage after desiccation and subsequent rehydration.

4) Some LEA proteins have been found to accumulate in plants during cold acclimation. Therefore, LEA proteins might participate in cryoprotection mechanisms, by accumulation in the tissues where primary ice nucleation occurs and act as stabilizers of glassy states (Wolkers et al. 2001). In various plant tissues, LEA proteins were able to prevent the inactivation of cyclic frozen and unfrozen lactate dehydrogenase (LDH) (Houde et al. 1995). Different hydrophilins, including LEA proteins from groups 2, 3, and 4, were able to prevent the inactivation of enzymes such as LDH or malate dehydrogenase (MDH) upon different dehydration levels (Goyal et al. 2005a, b; Reyes et al. 2005, 2008). Similarly, protective properties were also detected during freeze-thaw *in vitro* assays (Hara et al. 1999; Reyes et al. 2008). Interestingly, in these latter assays, a particular protective role was detected for the K-segment, a LEA2 protein distinctive motif, and for a conserved region in LEA4 proteins (Reyes et al. 2008). Because in these assays, 1:1 ratios between LEA and target proteins were enough to provide protection, it was proposed that this protective activity is carried out by protein-protein interactions (Reyes et al. 2005; Olvera-Carrillo et al. 2011).

5) LEA proteins can also act as molecular chaperones or shields that prevent irreversible protein aggregation during stress conditions (Goyal et al. 2005a, b; Chakrabortee et al. 2007); this finding proves

that LEA proteins can bind to non native proteins to prevent protein aggregation and maintain them in a folding-competent state. Dehydrins provide a physical interface between exposed hydrophobic patches of partially denatured proteins and the stabilizing influences of the preferential exclusion mechanism. The working hypothesis is that dehydrins are solubilizing agents with detergent and chaperone properties.

6) Exposure to adverse conditions often causes oxidative stress and reactive oxygen species (ROS) generation in plants. Therefore, an effective ROS-scavenging system is important for plants. LEA proteins can directly reduce oxidative stress by scavenging ROS and indirectly reduce ROS production by sequestering the toxic metal ions that generate ROS in dehydrating cells. During osmotic stress, dehydrins can act as antioxidants, scavenging the toxic quantities of metal ions in whole plants (Hara et al. 2005). Dehydrins scavenged the hydroxyl radical and peroxyl radical, but not the superoxide anion and hydrogen peroxide. *In vitro* experiments using a citrus dehydrin and GmPM1 or GmPM9 indicated that these proteins reduce the levels of hydroxyl radical (Hara et al. 2005; Roychoudhury et al. 2007; Liu et al. 2011). Several residues such as Lys, His, Gly and Ser are probably related to the radical scavenging because the residues are modified when the dehydrin scavenges the hydroxyl radical. LEA 2 proteins also have high affinity to several heavy metal ions such as $Fe^{3+}$, $Cu^{2+}$, $Zn^{2+}$, $Mn^{2+}$ and $Fe^{2+}$ (Herzer et al. 2003; Kruger et al. 2002; Svensson et al. 2000). The affinity might be caused by a high proportion of His residues in dehydrins. More recently, a dehydrin was shown to bind calcium in a phosphorylation-regulated mode (Alsheikh et al. 2003), and another LEA protein was assigned a role in iron transport in the phloem of *Ricinus communis* (Kruger et al. 2002).

7) Dehydrin-like proteins may also have role similar to compatible solutes (such as proline, sucrose and glycine betaine) involved in osmotic adjustment. A major problem under severe dehydration is that the loss of water leads to crystallization of cellular components, which in consequence damages cellular structures. This may be counteracted by LEA proteins, and some of the LEA proteins could essentially be considered as compatible solutes, which supports the likely role of sugars in maintaining the structure of the cytoplasm in the absence of water. Furthermore, the LEA proteins could be superior to sucrose as protectants in being less likely to crystallize. Some reports suggest

that the presence of sugars can enhance the protective effect of LEA proteins under dehydration (Black et al. 1999; Wolkers et al. 2001; Liu et al. 2010).

8) The LEA proteins act as nuclear proteins that unwind or repair DNA, regulate transcription, and might be associated with chromatin or cytoskeleton (Wise and Tunnaclife 2004). Their nuclear localization has also been associated with binding to DNA for some LEA2, LEA4, and LEA7 proteins (Colmenero-Flores et al. 1999; Maskin et al. 2007; Hara et al. 2009). Moreover, during germination, LEA proteins may regulate α-amylase activity and starch grain degradation throughout the time of water deficit (Rinne et al. 1999).

9) The mitochondrial localization of pea (*Pisum sativum*) LEA3 proteins correlates with their ability to protect some mitochondrial enzymes such as rhodanese and fumarase from inactivation induced by *in vitro* dehydration (Grelet et al. 2005). In a survey of pea mitochondrial proteome during development, a putative seed mitochondrial protein exhibited peptide tag sequence similarities with a soybean protein annotated in databases as a LEA-like protein (Bardel et al. 2002). The heat-soluble proteins cross reacting with dehydrin antibodies have been detected in mitochondrial fractions from cereals (Borovskii et al. 2000, 2002), but neither their primary sequence nor their subcellular localization has been demonstrated.

Even though all the above observations may be difficult to integrate, the wide variety of functions that have been attributed to these proteins could be related to their selection throughout evolution not only by low water availability, but also by other intrinsically associated conditions that would be the case of high temperatures or the generation of high ion concentrations or ROS. In addition, the flexibility of their structure should also be considered, accounting for their multi-functionality (Hara 2010; Liu et al. 2010, 2013; Olvera-Carrillo et al. 2011; Dominguez et al. 2013).


## 6. STRESS-REGULATED EXPRESSION OF LEA GENES/PROTEINS IN PLANTS

A list of some commonly expressed *LEA* genes is shown in Table 1. As discussed in the earlier sections, LEA protein or gene expression in terms of time course starts from the late period of maturation and reaches its peak during progressive dehydration, but sharply decreases after some hours of germination following water imbibition of seeds (Wang et al. 2007). Almost all of these *LEA* genes are expressed mostly in embryo or seeds, while some shows no tissue-specificity; hence expressed in cotyledons, panicles and also in stems, leaves and roots (vegetative tissues). The various factors and conditions influence LEA protein expression, among which the phytohormone abscisic acid (ABA) is considered the most important. Information from the model plant *Arabidopsis thaliana* has shown that there is a universal signal transduction network system in higher plants (Hundertmark and Hincha 2008). The endogenous ABA concentration enhances several folds as signal stimulus during any form of abiotic stress; for instance, in barley seedlings and beans, drought can result in high levels of ABA by 57–160 times compared with those of controls (Moons et al. 1995). At least, four steps are involved in LEA protein expression and regulation induced by stress: signal recognition, signal transduction, signal amplification and integration, LEA protein expression responses and its product formation. Concerning this aspect, there are at least two different pathways: ABA-dependent and ABA-independent types. ABA-dependent type gene expression is determined by endogenous ABA accumulation or exogenous ABA application. The maximum level of accumulation of *LEA* transcripts and LEA proteins coincide with the maximum accumulation of endogenous ABA. Many genes induced by ABA have been isolated, cloned, identified and subjected to functional analysis from plants, including *Physcomitrella patens*, *Pisum sativum* (Grelet et al. 2005), *Glycine max* (Shih et al. 2004), *Capsicum annuum* (Kim et al. 2005), *Oryza sativa* (Moons et al. 1997), *Gossypium hirsutum* (Galau et al. 1993) and *Raphanus sativus*. Examples of some of these genes include *Osem*, *SalT*, *Rab16* and *Rab25* from rice, *Em* and *Rab15* from wheat, *Rab17* and *Rab28* from maize, *Rab18* from *Arabidopsis*, *HVA1* and *HVA22* in barley, *le16*, *le4* and *TAS14* from tomato, *CDeT27-45* from *Craterostigma plantagineum*, dehydrins from maize and barley, and several other genes from cotton, rapeseed, soybean, carrot, potato etc. The *LEA* genes like *Em*, *Rab16A* and dehydrins in seeds can be found in the root, stem, leaf, callus and suspension cultures of higher plants under ABA or/and NaCl induction (Mundy and Chua 1988). In cotton seedlings, *in vitro* treatment could lead to the accumulation of LEA protein mRNAs.

**Table 1. Commonly expressed *LEA* genes from different plant species: Some earlier studies**

| Gene | Stress induced | Species | Organ Specificity | Reference |
| --- | --- | --- | --- | --- |
| *LEA D7, D11, D19, D29, D34, D113* | ? | Cotton | ? | Baker et al. (1988) |
| *Dehydrin* | D | Barley | --- | Close et al. (1989) |
| *pLEA76* | D | Rape | ? | Skriver and Mundy (1990) |
| *pMAH9* | D, W | Maize | --- | Gomez et al. (1988) |
| *Em* | D | Wheat | --- | Marcotte et al. (1988) |
| *salT* | O, D | Rice | stem | Claes et al. (1990) |
| *Rab16* | O, D, C | Rice | --- | Mundy and Chua (1988) |
| *Rab17* | O, D, C, W | Maize | --- | Vilardell et al. (1990) |
| *Rab15* | D | Wheat | root | King et al. (1992) |
| *Rab25* | O | Rice | ? | Kusano et al. (1992) |
| *Rab18* | C, D | *Arabidopsis* | --- | Lang and Palva (1992) |
| *Rab28* | D | Maize | --- | Niogret et al. (1996) |
| *Osem* | D | Rice | seed | Hattori et al. (1995) |
| *HVA1* | O, D, C, H | Barley | --- | Straub et al. (1994) |
| *HVA22* | D, C | Barley | --- | Shen et al. (2001) |
| *rd22* | O, D | *Arabidopsis* | --- | Yamaguchi-Shinozaki and |
| *rd29A* | O, D, C | *Arabidopsis* | --- | Shinozaki (1993a, b, 1994) |
| *rd29B* | O, D, C | *Arabidopsis* | leaf, petiole | Shinozaki (1993a, b, 1994) |
| *Dc-3, Dc-8* | D | Carrot | embryo | Dure et al. (1989) |
| *DcECP31* | D | Carrot | embryo | Ko and Kamada (2002) |

| Gene | Stress induced | Species | Organ Specificity | Reference |
| --- | --- | --- | --- | --- |
| *le16* | O, D, C, H | Tomato | leaf, petiole, stem, seed | Plant et al. (1991) |
| *Cor78/lti78* | D, C | *Arabidopsis* | --- | Horvath et al. (1993) |
| [*kin1, cor6.6/kin2, cor15a, cor47/rd17*] | O, D, C | *Arabidopsis* | ? | Thomashow (1998) |
| *CDeT27-45 Dehydrins B8, B9,* | D | *Craterostigma plantagineum* | --- | Bartels and Salamani (2001) |
| *rd19A, rd21A* | D | *Arabidopsis* | --- | Koizumi et al. (1993) |
| *RD26* | O, D | *Arabidopsis* | ? | Fujita et al. (2004) |
| *pPPC1* | D | Ice plant | --- | Vernon et al. (1993) |
| *Dehydrin M3* | D | Maize | --- | Close et al. (1989) |
| *Le4* | O, D, C, H | Tomato | shoot, seed | Swamy and Smith (1999) |
| *TAS14* | O, D | Tomato | --- | Godoy et al. (1990) |

O = high osmoticum (PEG or salt), D = dessication, C = cold, W = wound, H = heat, L= light, --- = denotes that not organ specific, ? = untested or unknown.

1) The Responsive to Abscisic acid (*Rab16A*) was the first novel *LEA* gene cloned from rice (Mundy and Chua 1988). Four tightly linked *Rab* genes (*Rab16A–D*) are differentially expressed in rice. The *Rab16* mRNA and protein were shown to accumulate in rice embryos, leaves, roots and callus-derived suspension cells upon treatment with NaCl and/or ABA. The genes *A* and *D* accumulate to higher levels in ABA-treated germinating seeds than *B* and *C*. In contrast to genes *A-C*, gene *D* does not accumulate to detectable levels in mature embryos, and hence shows no expression in developing seed tissues. Furthermore, ABA and osmotic stress differentially affect the accumulation in cultured suspension cells of the gene *D*, in contrast to their effects on the accumulation of *A-C* mRNA (Yamaguchi-Shinozaki et al. 1989). During seed development, low levels of *Rab16A* mRNA are found in both endosperm and embryo seed halves, whereas high levels are found in the resting grain or subjecting the root/shoot tissues to exogenous ABA/NaCl/ water stress. In suspension cultured cells, they are detected only after 15 min of ABA treatment, suggesting that protein synthesis is not required for *Rab16A* gene expression, but probably involves modification of pre-existing nuclear and/or cytosolic factors. The *Rab16A* gene was earlier found to be differentially regulated in salt-sensitive and salt-tolerant rice varieties, with the tolerant varieties (having higher endogenous ABA level) showing constitutive and higher level of expression than the salt-sensitive varieties, where the expression was induced only by salt treatment; the transcript or the protein being undetectable at the background level (Mukherjee et al. 2006; Roychoudhury et al. 2008). In case of seeds, only upon treatment with exogenous ABA could the gene expression be detected in salt-sensitive rice, while the seeds of salt-tolerant rice variety showed *Rab16A* gene or protein expression even in dry or water-imbibed conditions (Roychoudhury et al. 2009), proving again that ABA positively regulates *Rab* gene. A rice *Rab16* homologue gene, *Rab25* was obtained from a rice cDNA library prepared from ABA-treated suspension calli, showing 60.2% homology with rice *Rab16* (Kusano et al. 1992). Transcription of *Rab25* in rice calli was readily induced by either ABA (10 μM) treatment or high osmotic (0.5 M mannitol) stress. In maize, a homologue of *Rab16A,* called *Rab17* was identified (Vilardell et al. 1990). The transcript accumulated in mature embryos, disappeared during embryo germination, but the transcript detectable after 100 μM

ABA incubation of three-day old seedlings. The maize *Rab17* gene expression was also detected in all parts of seedlings; the induction was noted in both leaves and roots in response to 20 µM ABA. The leaves also responded to high NaCl concentration by increased level of *Rab17* mRNA with a very low induction in roots (Busk et al. 1997). The maize *Rab28* mRNA showed a similar pattern of accumulation as *Rab17* mRNA during embryogenesis. The mRNA levels were detected in embryos, 25 days after pollination (DAP) and increased as maturation proceeded. The mRNA levels also increased after ABA treatment. Dehydration stress also induced the transcript level in vegetative tissues. The gene expression in developing tissues and vascular structures indicated that Rab28 protein might be involved in late embryo differentiation processes (Niogret et al. 1996). An ABA-responsive gene *Rab15* was isolated from water stressed wheat roots (King et al. 1992). In *Arabidopsis, Rab18,* induced by ABA during cold acclimation process or water stress, was derived by screening a low-temperature-induced cDNA library with heterologous rice *Rab16A* cDNA probe (Lang and Palva 1992). Exogenous ABA could complement the reduced accumulation of *Rab18* mRNA to the wild type level in the ABA synthesis mutant, *abi-1*. In contrast, the expression of *Rab18* could be induced by exogenous ABA in *abi-3*, another ABA-response mutant. The amino acid sequence GGQ/GGYGT/S, repeated five times, in the N-terminal half, is unique to Rab18 protein. Three other conserved domains, DEYGNP near N-terminus, the imperfect Ser-rich repeat (RSGSGSSSSSEDDG) in the middle part and the imperfect Lys-rich repeats (GGRR/EKKGI/MT/MQ/DKIKEKLPG) in the C-terminal half are found almost identical in similar Rab proteins of other plant species.

2) Another multicopy *LEA* gene from rice, produced in large amounts, only in certain restricted plant parts is *salT*, mostly expressed in the root and sheath with temperature increase or excess salt or ABA, but least expressed in the lamina. The gene is also expressed in seeds, even in the absence of any stress imposed and developmentally regulated in young seedlings. The protein participates in developmental and global primary responses to salt stress, dehydration and phytohormones, thus preventing damage and rendering overall defense, helping the cells to become tolerant (Claes et al. 1990).

3) More than 100 *Em* genes have been identified from various species, including vascular and non-vascular plants, bacteria and brine shrimp. The most significant feature of Em proteins is a hydrophilic 20-mer motif of 1–4 repeats near the C-terminus in plants (Delseny et al. 2001) or 1–7 repeats in bacteria or brine shrimp. Although the 20-mer motifs among plants, bacteria and animals have limited sequence similarities, they share a common primary sequence, viz., GXKGGX13EM/I/L. The wheat *Em* gene is expressed in matured embryos and in vegetative tissues, in response to ABA/salt/osmotic stress. The accumulation of *Em* mRNA can be induced in isolated developing wheat embryos, in culture by ABA application, which prevents precocious germination (Morris et al. 1990). The induction of *Em* gene expression by osmotic stress both in immature and germinating embryos is consistent with the proposed roles of Em protein in mediating desiccation tolerance. The *Em* mRNA showed rapid accumulation within 60 min with the maximum accumulation after 12-24 h in rice suspension cultures, derived from immature rice embryos, following exposure of culture to 50 µM ABA or 0.4 M NaCl. The NaCl-induced *Em* expression was accompanied by a doubling of endogenous ABA levels. Desiccation of culture to 12-15% of their initial fresh weight also resulted in increase in the endogenous ABA levels and *Em* mRNA levels. The synergistic applications of saturating levels of both NaCl (0.1 M) and ABA (1 µM) doubled the *Em* transcript levels over maximum signal for each treatment alone (Bostock and Quatrano 1992). Shih et al. (2010) characterized a novel *Em*–like gene, *OsLEA1a* of rice. The encoded OsLEA1a protein has an N-terminal sequence similar to that of other plant Em proteins but lacks a 20-mer motif that is the most significant feature of typical Em proteins. The location of the sole intron indicates that the second exon of *OsLEA1a* is the mutated product of a typical *Em* gene. Transcriptome analysis revealed that *OsLEA1a* is mainly expressed in embryos, with no or only a few transcripts in osmotic stress-treated vegetative tissues. Structural analysis revealed that the OsLEA1a protein adopts high amounts of disordered conformations in solution and undergoes desiccation-induced conformational changes. Macromolecular interaction studies revealed that OsLEA1a protein interacts with non-reducing sugars and phospholipids but not poly-L-lysine. Thus, although the OsLEA1a protein lost its 20-mer motif, it is still involved in the formation of

bioglasses with non-reducing sugars or plasma membrane. However, the protein does not function as a chaperone as do other groups of hydrophilic LEA proteins. The orthologues of the *OsLEA1a* gene had been identified from various grasses but not in dicot plants. Genetic analysis indicated that rice *OsLEA1a* locates at a 193 Kbp segment in chromosome 1 and is conserved in several published cereal genomes. Thus, the ancestor of *Em*-like genes might have evolved after the divergence of monocot plants.

4) The rice *Osem* (homologous to wheat *Em*) gene is expressed in dry mature rice embryos and in early stages of germination, as well as in vegetative tissues in response to ABA. The *Osem* mRNA disappeared after 10 h imbibition of isolated dry embryos but was re-induced by further incubation with $10^{-5}$ M of ABA for 6 h. The expression of *Osem* was not induced in shoot tissues by treatment with ABA or desiccation (Hattori et al. 1995). The *Arabidopsis* mutant with a T-DNA insertion allele of the *AtEm6* gene was found to alternate the processes of seed dehydration and maturation, which suggests that the ATEM6 protein might be required for normal seed development and acquiring desiccation tolerance (Manfre et al. 2006, 2009).

5) The expression of barley *HVA1* is tissue specific and temporally regulated in developing seeds, with its mRNA detected in aleurone layers and embryos, but not in endosperm. In vegetative tissues, *HVA1* is also developmentally and organ specifically regulated by ABA and drought treatment. Accumulation of its mRNA is also induced by salt, cold and heat stress. The effect of salinity (0.17 M NaCl) and heat was much more modest compared to the other conditions. Heat induced the expression of *HVA1* only in roots but not in shoots. The effect of cold induction of *HVA1* could be readily reverted; no *HVA1* transcripts were detected after the cold-stressed plants were transferred back to normal temperature for 14 h (Straub et al. 1994). Another barley gene *HVA22* appears to be tissue specific with the level of its mRNA readily detectable in aleurone layers as early as 15 days after anthesis and remained high throughout the later stages. The mRNA was also abundant in embryos through all stages of seed development, yet undetectable in starchy endosperm. The expression of *HVA22* in vegetative tissues can be induced by ABA, drought and cold. The turnover rate of *HVA22* mRNA is correlated with the seed dormancy status; the mRNA degraded rapidly during imbibition but remained high even after 48 h of ABA (20 µM)

imbibition. In vegetative tissues, *HVA22* expression is regulated by ABA and environmental stresses. The transcript was induced by 100 μM ABA for 24 h. Dehydration and cold stress ($1°C$ for 4 days) also induced the gene expression in shoots and roots (Shen et al. 2001).

6) The accumulation of the group 3 *LEA* transcripts has been associated with genotypic differences in the acclimation to cold in rice varieties (Takahashi et al. 1994). Moons et al. (1995) identified a cDNA clone *oslea3*, encoding a group 3 LEA protein from roots of rice seedlings. The encoded OSLEA3 protein was found to accumulate to higher levels in roots of two salt-tolerant compared to a salt-sensitive rice variety in response to ABA. Exogenous application of ABA and exposure to salt shock (150 mM NaCl) rapidly induced *oslea3* transcript accumulation in seedling roots. The stress-induced *oslea3* transcript gradually declined upon prolonged salt shock, as wilting-induced damage became irreversible. The *oslea3* expression was compared for the salt-tolerant variety Pokkali and the salt-sensitive cultivar Taichung N1. Higher maximal mRNA levels were found in the roots of the tolerant variety, also declining less rapidly upon sustained salt shock, concomitant with a delayed drop in shoot water content. The results suggested that a differential regulation of *oslea3* expression is an aspect of the varietal differences in salt stress tolerance. Duan and Cai (2012) reported the isolation and functional characterization of the *OsLEA3-2* gene, which encodes a group 3 LEA protein in rice. A semi quantitative RT-PCR analysis revealed that the *OsLEA3-2* gene does not express in vegetative tissues under normal conditions; however, it can be induced by abiotic stresses by treating the seedlings with mannitol, salt or polyethylene glycol (PEG), but not with low temperature or ABA. The mannitol and salt stress also induced *OsLEA3-2* gene expression in the shoot base and leaves, while the PEG treatment induced gene expression in both roots and shoots. The *OsLEA3-2* gene was found to be located in the nucleus. A total of 34 rice LEA (*OsLEA*) genes were identified, of which 25 were new. Sixteen genes appeared to be up regulated by ABA. Six genes (*OsLEA3, 7, 21, 22, 27, and 28*) were up regulated by drought.

7) The molecular characterization and functional analysis of *OsLEA5* gene from rice was done by He et al. (2012). The OsLEA5 polypeptide is rich in Leu (10%), Ser (8.6%), and Asp (8.6%), while Cys, Trp, and Gln residue contents are very low, which are 2, 1.3 and

1.3%, respectively. Real-time PCR analysis showed that *OsLEA5* was expressed in different tissues during different developmental stages of rice. The expression levels of *OsLEA5* in the roots and panicles of full ripe stage were dramatically increased. The results of stress tolerance and cell viability assay demonstrated that recombinant *E. coli* cells producing OsLEA5 fusion protein exhibited improved resistance against diverse abiotic stresses like high salinity, osmotic, freezing or heat stress and UV radiation. The OsLEA5 protein conferred stabilization of the LDH under different abiotic stresses, such as heating, freeze-thawing and drying *in vitro*. The combined results indicated that OsLEA5 protein is a hydrophobic atypical LEA and closely associated with resistance to multiple abiotic stresses.

8) Reddy et al. (2012) characterized a cDNA clone encoding a LEA protein termed PgLEA from *Pennisetum glaucum* by screening a heat stress cDNA library. *PgLEA* cDNA encodes a 176 amino acid polypeptide with a predicted molecular mass of 19.21 kDa and an estimated isoelectric point of 7.77. The recombinant PgLEA protein expressed in *E. coli* possessed *in vitro* chaperone activity and protected the PgLEA-overproducing bacteria from damage caused by heat and salinity. Positive correlation existed between differentially up regulated *PgLEA* transcript levels and the duration and intensity of different environmental stresses. Transcript induction data, the presence of several putative stress-responsive transcription factor-binding sites in the promoter region of *PgLEA*, the *in vitro* chaperone activity of this protein and its protective effect against heat and salt damages in *E. coli* suggested their role in conferring abiotic stress tolerance in plants as well.

9) Colmenero-Flores et al. (1997) studied *Phaseolus vulgaris* by treatments of ABA and water stress; they noted induction of six cDNA clones, in which there were two types of *LEA* genes. Under the water pressure of 0.35 MPa (16 h), these genes reached their expression peak. In case of *Arachis hypogaea*, the expression of most *LEA* genes reached the peak level in 60 DAP seeds. Some peanut *LEA* genes, such as peanut *LEA2*, *LEA6* and *LEA7* showed similar expression patterns to their *Arabidopsis* counterpart. Expression of *AhLEA1*, *AhLEA4*, *AhLEA5* and *AhLEA8* genes was only detected in seed, but most of these genes in *Arabidopsis* expressed both in seed and non-seed tissues. The expression of all peanut *LEA3* genes was

detected in seed, while four of them were also detected in non-seed tissues. Interestingly, *AhLEA3-4* is found to be expressed not only in seed, but also in flower in high amounts, but not in other non-seed tissues (Su et al. 2004). Shih et al. (2004) characterized a LEA protein named GmPM16 from soybean (*Glycine max*) that interacts with sugar and forms tightly glassy matrixes in the dry state. The protein might play a role in reducing cellular damage in dry seeds by changing the protein conformation and forming tight cellular glasses. Savitri et al. (2013) also characterized a drought-tolerant gene *LEA-D-11* from soybean. Grelet et al. (2005) identified a LEA protein of group 3 (PsLEAm) that was localized within the matrix space of pea seed mitochondria. PsLEAm was expressed during late seed development and remained in the dry seed. Application of ABA was found to re-induce the expression of *PsLEAm* transcripts during germination. PsLEAm could not be detected in vegetative tissues; however, its expression could be re-induced in leaves by severe water stress. The recombinant PsLEAm was shown to protect two mitochondrial matrix enzymes, fumarase and rhodanase, during drying in an *in vitro* assay. This was the first characterization of a LEA protein in mitochondria and its beneficial role in desiccation tolerance. A *LEA* gene, *CarLEA4*, was isolated from chickpea, based on a cDNA library constructed with chickpea seedling leaves treated by PEG. The protein was localized in the nucleus. The transcripts of *CarLEA4* were detected in many chickpea organs including leaves, stems, roots, flowers, young pods and young seeds. *CarLEA4* was inhibited by leaf age and showed changes in expression pattern during seed development, pod development and germination. Furthermore, the expression of *CarLEA4* was strongly induced by drought, salt, heat, cold and ABA. CarLEA4 encodes a protein of LEA group 4 and involved in various plant developmental processes and abiotic stress responses (Gu et al. 2012). Recently, Battaglia and Covarrubias (2013) analyzed LEA proteins from those legumes whose complete genomes have been sequenced, such as *Phaseolus vulgaris, Glycine max, Medicago truncatula, Lotus japonicus, Cajanus cajan* and *Cicer arietinum*. Their sequence analysis allowed the recognition of novel legume specific motifs. *Medicago truncatula* microarrays showed that water deficit conditions imposed with NaCl treatments (200 mM) induce the accumulation of transcripts from *LEA* genes of groups 2, 3, 4, 6 and 7.

10) In case of *A. thaliana*, the gene *rd22* is induced by salt and water deficit, but not by cold or heat stress and also regulated by ABA. The *rd22* mRNA was also expressed during the early and middle stages of seed development, but not at the late stages. Moreover, protein synthesis is required for the induction of *rd22* mRNA, but not for *rd29* mRNA by ABA (Yamaguchi-Shinozaki and Shinozaki 1993a, b). The *rd29A* gene is induced within 20 min after dehydration began and was strongly expressed after 2 h, whereas *rd29B* was not detected until 2 h after dehydration. The maximum level of *rd29B* mRNA was detected at 10 h. Moreover, the *rd29A* mRNA was induced within 5 h after exposure to low temperature ($4^{\circ}$C) and was detectable for at least 24 h. The *rd29A* mRNA was detected within 1 h after high salt treatment and reached maximum at 5 h. By contrast, *rd29B* was induced slowly at 2 h after high salt treatment. Moreover, *rd29B* gene expression was dependent on ABA induction, while the induction of *rd29A* during cold stress was ABA-independent (Yamaguchi-Shinozaki and Shinozaki 1994). Singh et al. (2005) reported the three-dimensional structure of LEA protein from *A. thaliana* gene At1g01470.1. This protein has been classified as a LEA14 protein. The structure, which was determined by NMR spectroscopy, revealed that the At1g01470.1 protein has $\alpha\beta$-fold, consisting of one $\alpha$-helix and seven $\beta$-strands that form two antiparallel $\beta$-sheets. The closest structural homologues were discovered to be fibronectin Type III domains, which have <7% sequence identity. Because fibronectins from animal cells have been shown to be involved in cell adhesion, cell motility, wound healing and maintenance of cell shape, it is interesting to note that in plants, wounding or stress results in the overexpression of a protein with fibronectin Type III structural features.

11) Dalal et al. (2009) evaluated the functional role of a Group 4 LEA protein, LEA4-1 from *Brassica napus*. Expression analysis revealed that ABA, salt, cold and osmotic stresses induce the expression of *LEA4-1* gene in leaf tissues in *B. napus*. Conversely, reproductive tissues such as flowers and developing seeds showed constitutive expression of *LEA4*, which was up regulated in flowers under salt stress.

12) Kim et al. (2005) used differential screening to isolate a full-length dehydration-responsive cDNA clone encoding a hydrophobic LEA-like protein from PEG-treated hot pepper leaves. Named *CaLEA6* (for

*Capsicum annuum LEA*), this gene belongs to the atypical hydrophobic *LEA* Group 6. The full-length *CaLEA6* is 709 bp long with an open reading frame encoding 164 amino acids. It is predicted to produce a highly hydrophobic, but cytoplasmic protein. The putative molecular weight of CaLEA6 protein is 18 kDa, with a theoretical pI of 4.63. Based on Southern blot analysis, CaLEA6 appeared to exist as a small gene family. CaLEA6 was not expressed prior to any treatment, but its transcript was rapidly and greatly increased following trials with PEG, ABA and NaCl. Chilling also induced its rapid induction, but to a much lesser extent. Accumulation of CaLEA6 protein occurred soon after NaCl applications, but considerably delayed after treatment with PEG.

13) Zhao et al. (2010) cloned a member of the group 3 LEA, *MwLEA1*, from Mongolian wheatgrass (*Agropyron mongolium* Keng) based on a homologous sequence from wheat (*Triticum aestivum* L.). The protein had five repeat 11-amino-acid motifs, with a molecular weight of 19.4 kDa and a theoretical isoelectric point of 8.8. Subcellular localization indicated that the *MwLEA1* was localized in the nucleus of the onion epithelial cell. Under water stress conditions, *MwLEA1* exhibited different expression levels, which was higher in root and shoot but lowest in leaf. The expression profiling under different stresses indicated that *MwLEA1* played roles in responses to water, salt stresses as well as ABA regulation.

14) Du et al. (2013) identified 30 *LEA* genes from Chinese plum (*Prunus mume*) through genome-wide analysis. The *PmLEA* genes could be divided into eight groups (LEA_1, LEA_2, LEA_3, LEA_4, LEA_5, PvLEA18, dehydrin and seed maturation protein). Most *PmLEA* genes were highly expressed in flower and up regulated by ABA treatment. The expression of *PmLEA29* is the highest in all five organs (flower, fruit, leaf, root and stem), followed by *PmLEA30*, *PmLEA9* and *PmLEA13*. The expressions of most *PmLEA* genes were the highest in flower. Three genes (*PmLEA7*, *PmLEA10* and *PmLEA14*) showed strongest expression in stem, while *PmLEA3* and *PmLEA23* being highest in fruit, *PmLEA19* and *PmLEA16* highest in root. Generally, the expressions of dehydrin and LEA_2 members are higher than that of other groups.

George et al. (2009) reported the isolation and characterization of the cDNA clone for an atypical LEA protein (PjLEA3) from *Prosopis juliflora*. Unlike typical LEA proteins, rich in Gly, PjLEA3 has Ala as the most abundant amino acid followed by Ser and shows an average negative hydropathy. Northern analysis for *PjLEA3* in *P. juliflora* leaves under 90 mM $H_2O_2$ stress revealed up regulation of the transcript at 24 and 48 h.

15) Gao et al. (2012) characterized a *LEA* gene (*TaLEA1*) from *Tamarix androssowii* in response to heavy metal stress. They studied the *TaLEA1* expression in response to NaCl, $ZnCl_2$, $CuSO_4$ and $CdCl_2$. The gene was expressed in both roots and leaves, and the expressions in roots and leaves were differently regulated by the stressors. NaCl stress induced high expression levels of *TaLEA1* in leaves after 72 h of NaCl stress, the *TaLEA1* expression was 10.8-fold up regulated. However, NaCl stress did not induce considerable differences in the *TaLEA1* expression in roots.

ZnCl_2 stress induced high expression levels of *TaLEA1* in leaves, with the maximum expression (16.4-fold) after 6 h of stress; however, $ZnCl_2$ stress did not cause considerable differences in *TaLEA1* expression in roots. $CuSO_4$ stress did not cause considerable differences in the *TaLEA1* expression in roots. In leaves, $CuSO_4$ transiently inhibited *TaLEA1* expression at 6 and 24 h after induction of stress; however, *TaLEA1* was highly up regulated at 48 h and 72 h of $CuSO_4$ stress. Under $CdCl_2$ stress, the *TaLEA1* expression in roots was not differently regulated at 6 and 24 h of stress, but was up regulated at 48 h and 72 h of stress. In leaves, the *TaLEA1* expression was not differently regulated after 6 h of $CdCl_2$ stress, but was highly up regulated after 24–72 h of stress, with the expression peak (18.5-fold) after 48 h of stress.

16) The tomato *le16* is expressed in drought-stressed leaf, petiole and stem tissue and to a much lower extent in the pericarp of mature green tomato, developing seeds and non-stressed leaf tissue. The gene expression is induced by drought stress and elevated levels of endogenous ABA and by PEG-mediated water deficit, salinity, cold stress and heat stress in detached leaf tissues. The polypeptide is rich in Leu, Gly and Ala and containing both hydrophilic and hydrophobic domain. Another ABA-inducible, single copy gene from tomato, *TAS14*, is strongly induced by NaCl and ABA in tomato seedlings as well as in roots, stems and leaves; also by mannitol in seedlings. However, the gene is not induced by cold (Roychoudhury and Paul 2012).

17) A putative salt-tolerance gene *TsLEA1* was identified from the halophytic genetic model plant *Thellungiella salsuginea*, the protein showing sequence similarity to *Arabidopsis* LEA group 4 proteins. The transcription level of *TsLEA1* gene in *T. salsuginea* seedlings increased upon salt treatment and its transcript accumulated more in roots than in aerial parts (Zhang et al. 2012).

18) The *Craterostigma plantagineum pcC37-31* cDNA encodes the dsp-22 protein, whose mRNA levels increase in response to various stresses. The cDNA shows significant homology to early light-inducible protein (ELIP) genes. Light is involved in the regulation of the gene expression, and the encoded dsp-22 protein is chloroplastic. The ELIPs may play a role in the assembly of the photosystem. During desiccation, *C. plantagineum* chloroplasts undergo morphological changes, and thus the dsp-22 protein could bind pigments or help maintain assembled photosynthetic structures essential for resuming active photosynthesis during resurrection (Ingram and Bartels 1996). Among the other LEA proteins in *C. plantagineum* are pcC27-04 and pcC6-19 (D11-LEA-protein related), pcC3-06 (D7-LEA-protein related) and pcC 27-45 (D95-LEA-protein-related).

19) Even in case of lower plants like the moss *Physcomitrella patens*, the dehydration stress induced oscillatory increases in the steady-state levels of two Group 3-LEA proteins (Phypa_166566 and Phypa_211998). The amplitudes of the oscillatory increases are reflective of the severity of the dehydration stress. Dehydration stress increased ABA levels in *P. patens* and that ABA can also induce dosage-dependent oscillatory increases in steady-state transcript levels of these Group 3 *LEA* protein genes. These data suggest that, at least in the context of the moss, dehydration-induced temporal dynamics in steady-state transcript levels that is responsive to the severity of the stress may contribute to survival under periodic fluctuations of water availability and confer evolutionary advantages during colonization of land by plants. In addition to its ability to survive moderate dehydration stress, *P. patens* has also been shown to be highly tolerant to salt and osmotic stress, as *P. patens* can survive exposure to 350 mM NaCl and 500 mM sorbitol. salt- and osmotic-stress also induced dosage-dependent oscillatory increases in steady-state transcript levels of Group 3 *LEA* proteins, suggesting that temporal dynamics in steady-state transcript levels of abiotic stress-induced genes may be a

general phenomenon in *P. patens* and that these oscillatory increases may contribute to abiotic stress tolerance. The Phypa_211998 was shown to be up regulated under osmotic stress and not salt stress (Shinde et al. 2013).

20) The role of plant dehydrins in cold acclimation is well documented, which is highlighted in this section. Of the *Arabidopsis* dehydrin genes, *COR47* (Iwasaki et al. 1997), *Rab18* (Lang and Palva 1992), *Lti29* (*ERD10*) (Wellin et al. 1995), *Lti30* (*DHNXero2*) (Wellin et al. 1994), and *ERD14* (Kiyosue et al. 1994) have been reported to being up regulated under cold stress. Several cold-regulated (COR) proteins were also demonstrated to be similar to LEA 2 proteins. Alsheikh et al. (2003) found that acidic dehydrin ERD14 undergoes phosphorylation of several Ser residues upon cold, which is mediated by cold-regulated kinases. Phosphorylated ERD14 possesses a calcium-binding activity and the cytosolic concentration of calcium increase by several orders during stress. Calcium then binds to several specific proteins like calmodulin which then alters the activity of other proteins. Thus, ERD14 phosphorylation and its Ca-binding activity seem to be specifically induced by cold stress. The ERD14 possessing bound Ca may have a function of ionic buffer or sugar chaperone under cold stress. Alsheikh et al. (2005) showed that *in vitro* phosphorylated COR47 and ERD10 are also able to bind Ca and therefore it can be proposed that a Ca-binding activity is a trait shared by acidic dehydrins in *A. thaliana*. In *B. napus* and *B. juncea*, dehydrin genes named *BnDHN1* and *BjDHN1* were identified by Yao et al. (2005). Both the genes encode $Y_3SK_2$ dehydrins and expressed only in germinating seeds and that they enhance the cold tolerance during seedling emergence. Surprisingly, no *BnDHN1* or *BjDHN1* mRNAs were detected in dry seeds. In *Solanum tuberosum*, Kirch et al. (1997) have identified a stress-induced dehydrin gene *ci7*. Its expression is induced by cold (4°C), drought, high salinity and exogenous ABA. It is notable that the protein was detected only in tubers upon stress treatments while it was absent in leaves under the same conditions. In the wild potato (*Solanum sogarandinum*), Rorat et al. (2006) detected significant levels of DHN24 in transporting tissues, in apical parts, and in tubers under normal growth conditions, whereas no DHN24 was detected in leaves. Additionally, in *S. tuberosum* and *S. sogarandinum*, a KS-type dehydrin named DHN10 was detected in significant amounts in tubers, stems and flowers of

non-stressed plants (Rorat et al. 2004). The abundance of DHN10 depends on organ type and age. During low temperature (4°C) treatment, the DHN24 protein content substantially increased in tubers, in transporting organs and in apical parts, and only a small increase was observed in leaves. Contrary to DHN24, the amount of DHN10 increased in mature leaves under cold conditions. The increase in protein abundance (both DHN24 and DHN10) was observed only in the plants that were able to cold acclimate and it correlated with their acclimation capacity. These results suggested that the expression of both *Dhn24* and *Dhn10* are regulated by organ-specific factors under control conditions and by both organ specific and stress factors in mutual collaboration under stress conditions. In freezing-tolerant *S. commersonii* and freezing-sensitive *S. tuberosum* cv. Bintje, two homologous dehydrin genes *Scdhn1* and *Stdhn1*, have been identified by Baudo et al. (1996). It was demonstrated by the investigators that they are expressed in response to cold and ABA. In *Spinacia oleracea*, a cold-induced dehydrin CAP85 was identified by Neven et al. (1993). It has 11 copies of the K-segment within its molecule and exhibits a significant cryoprotective activity. In *Medicago sativa*, a dehydrin named CAS15 is associated with enhanced hardening capacity in response to cold (Monroy et al. 1993). Similarly, in cell suspension cultures of *M. falcata*, a dehydrin CAS18 was identified by Wolfraim et al. (1993) upon cold treatment. In *Cicer pinnatifidum*, a wild relative of important tropical and subtropical crop *C. arietinum*, a dehydrin gene named *cpdhn1* was identified by Bhattarai and Fettig (2005). The dehydrin protein, CpDHN1, accumulates in seeds during their maturation and it was also detected within leaves in response to drought, chilling (4°C), salinity and ABA. In *Vigna unguiculata*, an extremely chilling-sensitive annual crop, a 35 kDa protein enables young seedlings to emerge successfully under cold conditions in the field. In two closely related *V. unguiculata* lines (F6 siblines), the chilling-tolerant line containing this dehydrin showed maximal percentage emergence and slower leakage of electrolytes during seed imbibition than the genetically similar line without the dehydrin. The relation between physiological and molecular data suggested that this gene might be a useful marker in breeding programs (Ismail et al. 1997, 1999a, b). The 35-kDa protein present in the seeds of the cold-tolerant line 1393-2-11 was purified and described as DHN1. It was shown that its presence in the mature seeds of *V. unguiculata* co-segregated with chilling tolerance during seedling emergence. Two major groups of dehydrin genes induced by cold have been detected in wheat, the *WCS120* and the *WCOR410* (Fowler et al. 2001). During cold acclimation, the WCS120 proteins accumulate predominantly in the meristematic tissues because the survival of these tissues is crucial for the survival of the whole plant in the winter. The WCS120 protein possesses a relatively high cryoprotective activity in protecting the enzymatic activity of LDH. Therefore, the WCS120 protein acts as an important protective agent of many vital cellular proteins in cold-acclimated plant tissue (Sarhan et al. 1997). The WCOR410 proteins are highly hydrophilic, acidic dehydrins of the $SK_3$ type which have been found to be localized near the plasmalemma (Danyluk et al. 1998). In response to cold (2-4°C), the expression of DHN5 was detected in barley. Apart from *Dhn5*, the expression of *Dhn8* (an acidic $SK_3$ dehydrin, homologue of wheat WCOR410) has been reported by Zhu et al. (2000) under cold conditions (4°C), but was weaker when compared to *Dhn5*. Another small KS-type dehydrin, *Dhn13*, was found by Rodriguez et al. (2005) whose expression is constitutive, although it increases significantly upon abiotic stress conditions (2.8-fold upon cold and 8.5-fold upon mild sub-zero temperatures). Lee et al. (2005) characterized a full length cDNA of *OsDhn1* gene in rice that encodes acidic dehydrin family of hydrophilic protein (like the *Arabidopsis* COR47 and wheat WCOR410) having three K-type and one S-type motifs. The gene was strongly induced by low temperature and drought, and was regulated by CBF/DREB signaling pathway.

Accumulation of dehydrins also plays an important role in the acclimation of woody plants to unfavorable temperatures. In citrus trees, the dehydrins were first identified by Cai et al. (1995) in a cold-tolerant *Poncirus trifoliata*. The two cold-induced dehydrins of KS-type identified in *P. trifoliata* were described as COR11 and COR19. In *Citrus paradisi*, a dehydrin called COR15 was detected in peel tissue of mature fruits (Porat et al. 2002), whose expression enhances fruit chilling tolerance. A dehydrin named CuCOR19 was detected in the leaves of *C. unshiu* (Hara et al. 1999). Its expression was induced by cold (4°C) to significant levels, whereas increased concentrations of ABA (0.1 - 10 µM) or NaCl (50 - 200 mM) affected it only very slightly. Hara et al. (2001) showed a significant cryoprotective activity of CuCOR19 using catalase (CAT) and LDH assays. CuCOR19 also exhibited a radical-scavenging function against lipid peroxidation. Later, Hara et al. (2004) reported that CuCOR19 can scavenge hydroxyl radical and peroxyl radical. This protein is rich in Gly, His and Lys residues which are potential targets of these radicals. Hara et al. (2005) have also detected a new dehydrin in *C. unshiu* which was named CuCOR15. The authors found a significant metal-binding activity for this protein which is provided by its His-rich domains. The accumulation of CuCOR15 was enhanced by cold stress. The metal-binding activity of CuCOR15 is probably associated with its antioxidative activity, since free metal ions present an important catalytic agent for radical formation in the cells. Apart from these dehydrins, two dehydrin genes (*csDHN* and *cpDHN*) with the typical angiosperm-type K-segment were recently characterized by Porat et al. (2004) in *C. sinensis* and *C. paradisi*. Dehydrins have also been found in gymnosperm woody species. The gymnosperm K-segment consensus sequence is (Q/E)K(P/A)G(M/L) LDKIK(A/Q)(K/M)(I/L)PG, while the angiosperm K-segment consensus sequence is EKKGIMDKIKEKLPG (Close 1997). In two-year-old seedlings of *Pinus sylvestris*, a 60 kDa dehydrin was found by Kontunen-Soppela et al. (2000). The authors showed a decrease in the amount of this protein during seedling deacclimation in the spring. Nitrogen-fertilized seedlings showed a more rapid decrease in dehydrin content during de-hardening compared to control ones, since nitrogen fertilizers enhanced the renewed growth activity during de-hardening. In *Picea glauca*, Richard et al. (2000) characterized a dehydrin gene named *PgDhn1* isolated from a cDNA which was shown to encode a 27 kDa protein whose expression is induced by cold and drought treatments.

The mechanism of activation of LEA proteins during abiotic stress remains to be elucidated. A few studies have been conducted with respect to the Rab17 protein in maize (Figure 1). The Rab17 protein is the most heavily phosphorylated protein in the mature maize embryo. The predicted protein sequence of Rab17 contains a cluster of eight Ser residues (seven of them contiguous) in the central part, followed by a putative consensus site for CKII phosphorylation. It was shown that Rab17 is phosphorylated by CKII exclusively adjacent to the Ser residues. The Rab17 protein probably functions by interacting with specific proteins through association with their nuclear localization signal (NLS) peptides, either sequestering them in the cytoplasm and rendering them inactive through lack of access to their target sequences, or functioning as a nuclear export/import carrier. The binding with NLS was found to be dependent upon phosphorylation. The phosphorylation by CKII leads to the accumulation of the heavily phosphorylated protein during embryo maturation, both in the nucleus and cytoplasm. The phosphorylation states of

Rab17 were important for its optimal nuclear targeting, either by facilitating binding to specific proteins (co-transportation of the complex between Rab17 and other proteins) or as a dimer with higher efficiency forming a direct part of the nuclear targeting apparatus (Goday et al. 1994; Jensen et al. 1998). The lower level of phosphorylated Rab17 in callus cells might be due to the absence of specific kinase(s) or to inaccessibility of the protein substrate to the kinase(s) (Vilardell et al. 1990). The barley HVA22 protein also contains two consensus CKII phosphorylation sites (S/TXXD/E), with the first one SKVD located between positions 36 and 39, and the second between positions 54 and 57 (Shen et al. 2001).

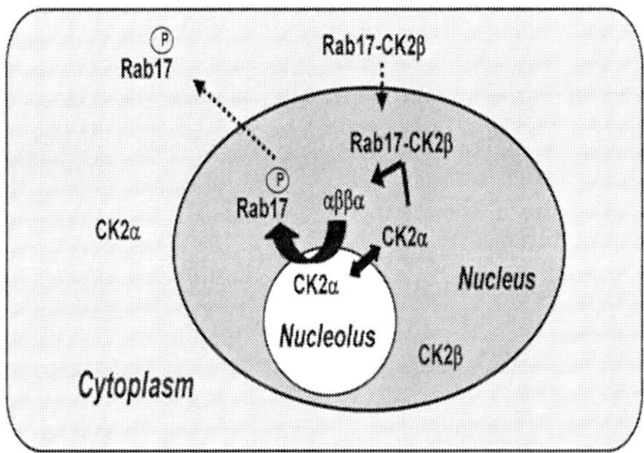


Figure 1. Model for nucleo-cytoplasmic trafficking of Rab17; mRab17 is retained in the nucleolus, whereas Rab17 is efficiently removed from this organelle. Phosphorylation of Rab17 could occur in the nucleus. The interaction between CK2β and Rab17 is relevant for nuclear targeting. After translation in the cytoplasm, Rab17 may interact with CK2β and the complex formed would travel to the nucleus where the three catalytic subunits of CK2 are predominantly located. In the nucleus, CK2α would disrupt Rab17/CK2β complex to associate with CK2β, generating the holoenzyme. Rab17 would probably be efficiently phosphorylated by CK2. Once phosphorylated, Rab17 would go to the nucleoplasm and/or cytoplasm to exert its function. [Figure modified from Jensen et al. (1998)].

By studying *A. thaliana* ABA-insensitive mutants (*aba* mutants), it was suggested that some of the *LEA* genes can be induced by drought, salt and cold; so it is considered that the expression of these genes do not need ABA under cold or drought. Several water stress-inducible *LEA* genes, such as *rd29A/low-temperature induced78/cold-regulated78* (*rd29A/lti78/cor78*), *rd17* (*cor47*), *cold-inducible1* (*kin1*), *erd10* and *cor6.6*, are known to be regulated by dehydration and cold stress in ABA-independent signaling pathways (Roychoudhury et al. 2013).

## 7. UPSTREAM ABIOTIC STRESS-INDUCIBLE ELEMENTS OF *LEA* GENES AND REGULATORY TRANS-ACTING FACTORS

It has been proved that cis-acting elements and trans-acting factors are related to ABA-responsive *LEA* gene transcriptional regulation. Sequence comparison and functional analysis of the promoters of the ABA-inducible *LEA* genes showed that one or more conserved, ABA-responsive, cis-acting element(s) called ABRE(s) are present. This element is defined as a conserved palindromic motif of 8-10 base pairs, viz., C/AACACGTGGCA, with the G-box ACGT as the core element (Busk and Pages 1998). The sequences flanking the hexameric G-box core (CACGTG) regulate the specificity of protein binding, determine tissue preference and affect *in vivo* regulation of gene expression (Roychoudhury and Paul 2012). The ABRE(s) function with a second sequence element that does not contain the ACGT core. Such GC-rich sequences called coupling elements (CE), usually sharing a CGCGTG consensus, are similar to ABREs, except that the A of the ACGT element is replaced by G in the latter. The two elements ABRE and CE, spaced at less than 20 bp distance, together constitute an ABA-responsive complex (ABRC), which can synergistically activate transcription by ABA (Busk and Pages 1998). It was also shown that ABRE(s) and CE(s) are functionally equivalent (Hobo et al. 1999). A comprehensive list of ABRE(s) and CE(s) in various *LEA* genes is already documented in details in one of our earlier communications (Roychoudhury and Paul 2012, and the references therein) and is beyond the scope of this chapter. In general, a single copy of ABRE seems insufficient for ABA-dependent transcription (Nakashima et al. 2009), and a minimal sequence unit necessary is composed of various combinations of ABREs or ABRE with CE. The multimerization of ABRE sequences or combinations of ABRE and CE from *LEA* genes found to function synergistically and producing enhancement in reporter gene activity, were used to create synthetic promoters (Cazzonelli and Velten 2008). For instance, the introgression of the two synthetically designed promoters, 4X ABRE (four tandem repeats from *Em*) and 2X ABRC (two copies of *Em* ABRE plus two copies of CE from *HVA22*), each fused with minimal promoter of CaMV35S promoter, could induce the *gus* expression in transgenic tobacco in response to high NaCl concentration, PEG-mediated water stress or ABA, but not at the constitutive level, proving that they are efficient stress-inducible promoters (Roychoudhury and Sengupta 2009). The above two promoters, together with *Rab16A* promoter, has also been shown to function, in a concentration-dependent manner, as potent ABA-inducible promoters in both vegetative and floral organs of transgenic rice (Ganguly et al. 2011). The basic leucine zipper (bZIP) group of transcription factors, defined by structural motif composed of a leucine zipper dimerization domain and a positively charged DNA interface (region of basic amino acids that directly contacts DNA adjacent to a hydrophobic heptad repeat), comprise the largest known family of ABA-inducible DNA-binding proteins that bind to ABRE(s) and/or CEs and effective in triggering the downstream *LEA* genes during salinity or drought stress. For instance, the nuclear factor OSBZ8 has been found to bind both motif I (ABRE) and motif II (CE-like motif) to transactivate *Rab16A* during salinity stress in rice (Roychoudhury et al. 2008). A list of such ABA-inducible trans-acting factors and their target genes are already presented earlier (Roychoudhury and Paul 2012). Most of these factors are post-translationally modified and activated through phosphorylation by several kinases (chiefly calcium-dependent) during stress response. In promoters such as *CDeT27–45* or *CDeT6–19,* isolated from *C. plantagineum,* G-box-related ABREs do not appear to be major determinants of ABA or drought response. Besides the ABA-mediated gene expression, the investigation of drought induced *LEA* genes in *A. thaliana* has also revealed ABA-independent signal transduction pathways. In the ABA-independent pathway, the C-repeat/dehydration response element (C/DRE), containing a core sequence of A/GCCGAC, has been shown to be essential for transcriptional activation in response to cold, drought, and/or high salt treatments (Yamaguchi Shinozaki and Shinozaki 1994). The *A. thaliana* genes *rd29A* and *rd29B* are differentially induced under conditions of dehydration, salt or cold stress, and ABA treatment. The *rd29A* gene has at least two cis-acting elements. The 9 bp direct repeat sequence (TACCGACAT) serving as the DRE, functions in the initial rapid response of *rd29A* to drought, salt or low temperature. The slower ABA response is mediated by another fragment that contains an ABRE. The DRE related motifs have also been reported in the promoter region of the cold-inducible *WCS120* gene from wheat. Subsequently, the CRT motif and the low-temperature responsive element (LTRE) were identified in the promoters of cold-regulated genes from *Arabidopsis* such as *kin1, kin2 and Rab18*, and were also responsible for the regulation of *BN115* and *WCS120* from *B. napus* and wheat, respectively (Roychoudhury et al. 2013). Wheat and barley possess a small family of *COR* genes including *WCS19* (Chauvin et al. 1993), *WCOR14* (Tsvetanov et al. 2000) and *BCOR14b* (Crosatti et al. 1999), all of which encode chloroplast-targeted COR proteins analogous to the *Arabidopsis* protein COR15a (Thomashow 1994). The wheat *COR* gene, *WCOR15*, which encodes a chloroplast-targeted protein and shares low-temperature specificity with the barley and wheat *COR* genes, has been isolated by Takumi et al. (2003). The expression of *WCOR15* was specifically induced by low temperature. The promoter of *WCOR15* contained at least three CRT/DRE-like sequence motifs found in *Arabidopsis COR* genes. The transcription factors that bind to the CRT/DRE have been isolated; CRT/DRE-binding factors (CBF or DREB), containing a DNA binding motif found in EREBP1 and AP2 transcriptional activators. However, maize DREBs like DBF1 and DBF2 are involved in *Rab17* gene regulation through two DREs in an ABA-dependent pathway (Kizis and Pages 2002).

## 8. OVEREXPRESSION OF *LEA* GENES TO GENERATE ABIOTIC STRESS-TOLERANT TRANSGENIC PLANTS

Table 2 shows the overexpression of *LEA* genes to generate abiotic stress-tolerant transgenic plants.

## CONCLUSION

LEA proteins have represented for a long time an enigma in plant biology. The lack of similarity to other proteins of known function and their high structural flexibility has hampered the progress regarding their activity and the implicated mechanisms. Of particular interest was the finding that various *LEA* transcripts from all groups accumulate in meristematic regions, even in plants grown under optimal growth conditions, supporting a protective role in these fragile but critical regions, for plant adaptation and survival. However, the high association between their accumulation and different levels of water deficit, caused during development or by the environment, has contributed to build hypotheses regarding their role in the plant tolerance to those conditions that produce low water availability, as drought, salinity or extreme temperatures. This has led to relevant advances concerning their transcript and protein accumulation and expression patterns in response to abiotic stress or throughout development, their participation in these processes, and their structural properties as described in this chapter. The available information presents LEA proteins as a set of proteins whose participation in plant tolerance to water deficit conditions has gained experimental support and also as proteins whose flexible structure turns them as a good model to find the functional relevance of this flexibility in the plant response to stress. The LEA proteins are multi-functional stress proteins helping to maintain normal metabolism for higher plants under severe conditions. The analysis of the compiled data from literature and from the available databases showed that most of the structural characteristics recognized in these proteins are conserved in their corresponding homologues in the different LEA protein groups of different plants.

The biophysical features of LEA proteins, as discussed in this chapter suggest that they may perform a bipartite function under different cellular water states. Thus, the different protective mechanisms may act at different stages of water loss. Moreover, because a number of non reducing oligosaccharides could be involved in drought and desiccation, the relation between LEA proteins and non reducing oligosaccharides and the role of oligosaccharides alone should not be ignored. Hence, considering the interaction between LEA proteins and oligosaccharides and the role of bioglasses in desiccation, LEA proteins should perform their functions in bioglasses in anhydrobiotic organisms. Under desiccation, the LEA proteins should also directly interact with proteins, oligosaccharides and the plasma membrane and may enhance bioglass strength, as well as act as a water replacement to stabilize cellular components.

Up to date, there have been mainly two ways to study LEA protein genes, viz., select mutant genes from model plants and analyze stress responses of crops. The extensive application of transgenic plants and microarrays is one developing direction. By utilizing an individual functional gene, the obtained transgenic plants can exhibit stress-tolerant characters, but the practical effect is not so good as that in production. So, transferring transcription factor genes (like *DREB* or *CBF* gene) into crops may be of greater benefit to bring out fine functions of LEA proteins and genes which will be the second promising direction in this field. Many results are gained under a given condition in laboratories and most of them at subcellular, cellular, tissue or organ levels. Higher plants in natural field suffer from many environmental stresses, hard to be controlled simultaneously by human being. Therefore, the results are inconsistent with, to some extent, to those in nature and at the whole plant level, which needs further investigation from whole-plant level to community level in the field environment.

**Table 2. Overexpression of *LEA* Genes to Generate Abiotic Stress-Tolerant Transgenic Plants**

| Gene | Source | Transgenic | Effect | Reference |
| --- | --- | --- | --- | --- |
| *Rab16A* | Rice (var. Pokkali) | i) Tobacco | The transgenic lines showed normal growth, morphology and seed production as the wild type (WT) plants without any yield penalty under stress conditions. They exhibited significantly increased tolerance to salinity (up to 300 mM NaCl), sustained growth rates under stress conditions; with concomitant increased osmolyte production like reducing sugars, proline and higher polyamines. They also showed delayed development of damaged symptoms with better antioxidative machinery and more favorable mineral balance, as reflected by reduced $H_2O_2$ levels and lipid peroxidation, lesser chlorophyll loss as well as lesser accumulation of $Na^+$ and greater accumulation of $K^+$ under 200 mM NaCl stress.<br><br>Enhanced tolerance to salinity. The superior physiological performances of the transgenics under salt treatment were reflected in lesser shoot or root length inhibition, reduced chlorophyll damages, lesser accumulation of $Na^+$ and reduced | Roychoudhury et al. (2007) |

| Gene | Source | Transgenic | Effect | Reference |
| --- | --- | --- | --- | --- |
| | | ii) Rice (var. Khitish) | loss of $K^+$, increased proline content as compared with the WT plants. All these results indicated that the overproduction of RAB16A protein in the transgenics enable them to display enhanced tolerance to salinity stress with improved physiological traits. | Ganguly et al. (2012) |
| *DHN-5* dehydrin | Wheat | *Arabidopsis* | Stronger growth under high concentrations of NaCl or under water deprivation, and a faster recovery from mannitol treatment. Leaf area and seed germination rate decreased much more in WT than in transgenic plants subjected to salt stress. Moreover, the water potential was more negative in transgenic than in WT plants. In addition, the transgenic plants had higher proline contents and lower water loss rate under water stress. An improved tolerance to salt and drought stress was achieved. | Brini et al. (2007) |
| *ZmDHN2b* dehydrin | Maize | Tobacco | Cold tolerance; accelerated seed germination and seedling growth at 15°C. Transgenic lines had lower levels of cold-induced malondialdehyde (MDA) and less electrolyte leakage than WT tobacco at 4°C. | Xing et al. (2012) |

**Table 2. (Continued)**

| Gene | Source | Transgenic | Effect | Reference |
|---|---|---|---|---|
| *TAS14* dehydrin | Tomato | Tomato | Improved long-term drought and salinity tolerance without affecting plant growth under non-stress conditions; osmotic stress tolerance via osmotic potential reduction and solute accumulation, such as sugars and $K^+$ in drought and salinity conditions; plants are able to distribute the $Na^+$ accumulation between young and adult leaves over a prolonged period in stressful conditions; an earlier and greater accumulation of ABA in leaves during short-term periods. | Muñoz-Mayor et al. (2012) |
| *Rab17* dehydrin | Maize | *Arabidopsis* | The constitutive expression produced high sugar and proline content; in addition, these plants showed more tolerance to high salinity conditions and recovered faster from mannitol treatment than non transformed control plants. | Figueras et al. (2004) |
| DHN24 | *Solanum sogarandinum* | Cucumber (*Cucumis sativus* cv. Borszagowski) | Enhanced freezing tolerance under cold stress (4°C) | Yin et al. (2006) |

| Gene | Source | Transgenic | Effect | Reference |
| --- | --- | --- | --- | --- |
| PMA1595 (Group I) or PMA80 (Group 2) | Wheat | Rice | Increased salt and drought stress tolerance | Cheng et al. (2002) |
| CuCOR19 | *Citrus unshiu* | Tobacco | Enhanced cold tolerance with no lipid peroxidation; the transgenic seeds germinated earlier under cold when compared to WT plants. | Hara et al. (2003) |
| WCOR410 (Group 2) | Wheat | Strawberry | Improved chilling tolerance at 5-8$^O$C after previous acclimation treatment | Houde et al. (2004) |
| Multiple dehydrins (*RAB18* and *COR47* combination) or (*LTI29* and *LTI30* combination) (double constructs) | *Arabidopsis* | *Arabidopsis* | Overexpression of the chimeric genes resulted in accumulation of the corresponding dehydrins to levels similar or higher than in cold-acclimated WT plants. Transgenic plants exhibited improved survival when exposed to freezing stress compared to the control plants | Puhakainen et al. (2004) |

**Table 2. (Continued)**

| Gene | Source | Transgenic | Effect | Reference |
|---|---|---|---|---|
| *Rab7* (encodes a GTP-binding protein involved in intracellular vesicle trafficking from late endosome to the vacuole | *Pennisetum glaucum* | Tobacco | Enhanced tolerance to NaCl and mannitol with increased alkaline phosphatase activity, showing that *Rab7* is a potential candidate gene for developing both salinity and dehydration tolerance *in planta*. | Agarwal et al. (2008) |
| *TsLEA-1* | *Thellungiella salsuginea* | *Arabidopsis* | Salt tolerance | Zhang et al. (2012) |
| *TaLEA-3* | Wheat | *Leymus chinensis* | Enhanced growth ability under drought stress, during which transgenic lines had increased the relative water content, leaf water potential, relative average growth rate, but decreased the MDA content compared with the non-transgenic plants. | Wang et al. (2009) |

| Gene | Source | Transgenic | Effect | Reference |
| --- | --- | --- | --- | --- |
| *WCS19* (*LEA3*) | Wheat | *Arabidopsis* | The gene encoding chloroplastic protein resulted in increased freezing tolerance in cold-acclimated plants | NDong et al. (2002) |
| *HVA1* | Barley | i) Mulberry (*Morus indica*) | Transgenic plants showed better cellular membrane stability, photosynthetic yield (PSII activity), less photo-oxidative damage and better water use efficiency as compared to the non-transgenic plants under both salinity and drought stress. Under salinity stress, transgenic plants showed many fold increase in proline concentration than the non-transgenic plants and under water deficit conditions, proline was accumulated only in the non-transgenic plants. The production of HVA1 proteins helped in better performance by protecting membrane stability of plasma membrane as well as chloroplastic membranes from injury under abiotic stress. Transgenic plants also exhibited tolerance against cold stress. Enhanced expression of stress | Lal et al. (2008); Checker et al. (2012) |

**Table 2. (Continued)**

| Gene | Source | Transgenic | Effect | Reference |
|---|---|---|---|---|
| | | | responsive genes such as *MidnaJ* and *Mi2-cysperoxidin* suggested that *HVA1* can regulate downstream genes associated with providing abiotic stress tolerance. | |
| | | ii) Creeping bentgrass (*Agrostis stolonifera* var. *palustris*), a drought-intolerant grass species | Lesser water-deficit injury by maintaining high water content in leaves and significantly less extent of leaf wilting compared with non-transgenic plants. | Fu et al. (2007) |
| | | iii) Common bean (*Phaseolus vulgaris*) | Drought tolerance with increase in root length of transgenic plants | Kwapata et al. (2012) |

| Gene | Source | Transgenic | Effect | Reference |
| --- | --- | --- | --- | --- |
| | | iv) Rice | Increased tolerance to water deficit and salinity (200 mM NaCl) with higher growth rates under stress conditions. | Xu et al. (1996) |
| | | | Under prolonged drought stress, the transgenic plants maintained higher leaf relative water content (RWC) and showed lesser reduction in plant growth, delaying wilting by more than two weeks as compared to WT plants. Transgenic lines had relatively better cell membrane protection than WT line, 28 days after stress. | Babu et al. (2004) |
| | | v) Wheat | Transgenic plants exhibited higher water use efficiency under water deficit conditions, along with greater total dry mass, root fresh and dry weights, and shoot dry weight, thus showing improved growth characteristics. | Sivamani et al. (2000) |
| *AtLEA3-3* | *Arabidopsis* | *Arabidopsis* | Salt and osmotic stress tolerance that is characterized during germination and early seedling establishment | Zhao et al. (2011) |
| *OsLEA3-1* | Rice (var. IRAT109) | Rice (drought-sensitive rice Zhonghua 11) | Higher grain yield than the WT under drought stress; enhanced drought resistance without yield penalty | Xiao et al. (2007) |

**Table 2. (Continued)**

| Gene | Source | Transgenic | Effect | Reference |
|---|---|---|---|---|
| *OsLEA3-1* | Rice | *Arabidopsis* and rice | Transgenic *Arabidopsis* seedlings showed better growth on media supplemented with 150 mM mannitol or 100 mM NaCl as compared with WT plants. The transgenic rice also showed significantly stronger growth performance than control under salinity or osmotic stress conditions and was able to recover after 20 days of drought stress. | Duan and Cai (2012) |
| *LEA4-1* | *Brassica napus* | *Arabidopsis* | Enhanced tolerance to salt and drought stresses | Dalal et al. (2009) |
| *Rab28* (Group 5 LEA) | Maize | Maize | Transgenic plants showed sustained growth under PEG-mediated dehydration compared to WT controls. Under osmotic stress, transgenic seedlings showed increased leaf and root area, higher relative water content, reduced chlorophyll loss and lower MDA production in relation to WT plants. Moreover, transgenic seeds exhibited higher germination rates than WT seeds under water deficit. The Rab28 protein was detected in nucleolar structures in transgenic plants. | Amara et al. (2013) |
| *CaLEA6* (Group 5) | Hot pepper | Tobacco | Enhanced tolerance to dehydration and salt stresses, but not chilling | Kim et al. (2005) |

| Gene | Source | Transgenic | Effect | Reference |
| --- | --- | --- | --- | --- |
| *AtLEA6-1* (previously called Sag21 or AtLEA5) | *Arabidopsis* | *Arabidopsis* | Increased the root growth and shoot biomass, both in optimal condition and under $H_2O_2$ stress. However, the photosynthesis of transgenic plants was more susceptible to drought | Miller et al. (1999); Mowla et al. (2006) |
| *TaLEA1* | *Tamarix androssowii* | Poplar | The transgenic plants showed better growth than the WT plants, indicating that *TaLEA1* provides tolerance to $CdCl_2$ stress. These results suggested that *TaLEA1* confers tolerance to $CdCl_2$ stress by enhancing ROS-scavenging ability (higher superoxide dismutase and peroxidase activities) and decreased lipid peroxidation (lower MDA levels). | Gao et al. (2013) |

## ACKNOWLEDGMENTS

Financial grant from the Science and Engineering Research Board, Department of Science and Technology, Government of India (SR/FT/LS-65/2010) to Dr. Aryadeep Roychoudhury is gratefully acknowledged.

## REFERENCES

Agarwal PK, Agarwal P, Jain P, Jha B, Reddy MK, Sopory SK (2008) Constitutive overexpression of a stress-inducible small GTP-binding protein PgRab7 from *Pennisetum glaucum* enhances abiotic stress tolerance in transgenic tobacco. *Plant Cell Rep* 27: 105-115.

Alsheikh MK, Heyen BJ, Randall SK (2003) Ion binding properties of the dehydrin ERD14 are dependent upon phosphorylation. *J Biol Chem* 278: 40882–40889.

Alsheikh MK, Svensson JT, Randall SK (2005) Phosphorylation regulated ion-binding is a property shared by the acidic subclass dehydrins. *Plant Cell Environ* 28: 1114-1122.

Amara I, Capellades M, Ludevid MD, Pagès M, Goday A (2013) Enhanced water stress tolerance of transgenic maize plants over-expressing *LEA Rab28* gene. *J Plant Physiol* 170: 864-873.

Ashghar R, Fenton RD, Demason DA, Close TJ (1994) Nuclear and cytoplasmic localization of maize embryo and aleurone dehydrin. *Protoplasma* 177: 87-94.

Babu RC, Zhang J, Blum A, Ho T-HD, Wu R, Nguyen HT (2004) *HVA1*, a *LEA* gene from barley confers dehydration tolerance in transgenic rice (*Oryza sativa* L.) via cell membrane protection. *Plant Sci* 166: 855–862.

Bahrndorff S, Tunnacliffe A, Wise MJ, McGee B, Holmstrup M, Loeschcke V (2009) Bioinformatics and protein expression analyses implicate LEA proteins in the drought response of Collembola. *J Insect Physiol* 55: 210-217.

Baker J, Steel C, Dure L, III (1988) Sequence and characterization of 6 LEA proteins and their genes from cotton. *Plant Mol Biol* 11: 277–291.

Bardel J, Louwagie M, Jaquinod M, Jourdain A, Luche S, Rabilloud T, et al. (2002) A survey of the plant mitochondrial proteome in relation to development. *Proteomics* 2: 880–898.

Bartels D, Salamani F (2001) Desiccation tolerance in the resurrection plant *Craterostigma plantagineum*. A contribution to the study of drought tolerance at the molecular level. *Plant Physiol* 127: 1346-1353.

Battaglia M, Covarrubias AA (2013) Late Embryogenesis Abundant (LEA) proteins in legumes *Front Plant Sci* 4: 190.

Battaglia M, Olvera-Carrillo Y, Garciarrubio A, Campos F, Covarrubias AA (2008) The enigmatic LEA proteins and other hydrophilins. *Plant Physiol* 148: 6–24.

Baudo MM, Meza-Zepeda LA, Palva ET, Heino P (1996) Induction of homologous low temperature and ABA responsive genes in frost resistant (*Solanum commersonii*) and frost sensitive (*Solanum tuberosum* cv. Bintje) potato species. *Plant Mol Biol* 30: 331-336.

Bhattarai T, Fettig S (2005) Isolation and characterization of a dehydrin gene from *Cicer pinnatifidum*, a drought-resistant wild relative of chickpea. *Physiol Plant* 123: 452-458.

Bies-Etheve N, Gaubier-Comella P, Debures A, Lasserre E, Jobet E, Raynal M, et al. (2008) Inventory, evolution and expression profiling diversity of the LEA (late embryogenesis abundant) protein gene family in *Arabidopsis thaliana. Plant Mol Biol* 67: 107–124.

Black M, Corbineau F, Gee H, Come D (1999) Water content, raffinose, and dehydrins in the induction of desiccation tolerance in immature wheat embryos. *Plant Physiol* 120: 463–472.

Blackman SA, Obendorf RL, Leopold AC (1995) Desiccation tolerance in developing soybean seeds: The role of stress proteins. *Physiol Plant* 93: 630–638.

Borovskii GB, Stupnikova IV, Antipina AI, Downs CA, Voinikov VK (2000) Accumulation of dehydrin-like proteins in the mitochondria of cold-treated plants. *J Plant Physiol* 156: 797–800.

Borovskii GB, Stupnikova IV, Antipina AI, Vladimirova SV, Voinikov VK (2002) Accumulation of dehydrin-like proteins in the mitochondria of cereals in response to cold, freezing, drought and ABA treatment. *BMC Plant Biol* 2: 1–7.

Bostock RM, Quatrano RS (1992) Regulation of *Em* gene expression in rice. *Plant Physiol* 98: 1356-1363.

Bray EAJ, Bailey-Serres E, Weretilnyk E (2000) Responses to abiotic stresses, In: Biochemistry and Molecular Biology of Plants. Buchanan W, Gruissem R Jones (Eds.), *American Society of Plant Physiologists*, pp. 1158–1176.

Breton G, Danyluk J, Charron JB, Sarhan F (2003) Expression profiling and bioinformatic analyses of a novel stress-regulated multispanning transmembrane protein family from cereals and *Arabidopsis*. *Plant Physiol* 132: 64–74.

Brini F, Hanin M, Lumbreras V, Amara I, Khoudi H, Hassairi A, et al. (2007) Overexpression of wheat dehydrin DHN-5 enhances tolerance to salt and osmotic stress in *Arabidopsis thaliana*. *Plant Cell Rep* 26: 2017-2026.

Brini F, Yamamoto A, Jlaiel L, Takeda S, Hobo T, Dinh HQ, et al. (2011) Pleiotropic effects of the wheat dehydrin DHN-5 on stress responses in *Arabidopsis*. *Plant Cell Physiol* 52: 676-688.

Buitink J, Leprince O (2004) Glass formation in plant anhydrobiotes: Survival in the dry state. *Cryobiology* 48: 215–228.

Busk PK, Jensen AB, Pages M (1997) Regulatory elements in vivo in the promoter of the abscisic acid responsive gene *rab17* from maize. *Plant J* 11: 1285-1295.

Busk PK, Pages M (1998) Regulation of abscisic acid-induced transcription. *Plant Mol Biol* 37: 425-435.

Cai Q, Moore GA, Guy CL (1995) An unusual group 2 LEA gene family in citrus responsive to low temperature. *Plant Mol Biol* 29: 11-23.

Campbell SA, Close TJ (1997) Dehydrins: genes, proteins, and associations with phenotypic traits. *New Phytol* 137: 61–74.

Caruso A, Morabito D, Delmotte F, Kahlem G, Carpin S (2002) Dehydrin induction during drought and osmotic stress in *Populus*. *Plant Physiol Biochem* 40: 1033-1042.

Cazzonelli CI, Velten J (2008) In vivo characterization of plant promoter element interaction using synthetic promoters. *Transgenic Res* 17: 437–457.

Chakrabortee S, Boschetti C, Walton LJ, Sarkar S, Rubinsztein DC, Tunnacliffe A (2007) Hydrophilic protein associated with desiccation tolerance exhibits broad protein stabilization function. *Proc Natl Acad Sci USA* 104: 18073–18078.

Chauvin LP, Houde M, Sarhan F (1993) A leaf-specific gene stimulated by light during wheat acclimation to low temperature. *Plant Mol Biol* 23: 255-265.

Chaves MM, Maroco JP, Pereira JS (2003) Understanding plant responses to drought—from genes to the whole plant, *Funct Plant Biol* 30: 239–264.

Checker VG, Chhibbar AK, Khurana P (2012) Stress-inducible expression of barley *Hva1* gene in transgenic mulberry displays enhanced tolerance against drought, salinity and cold stress. *Transgenic Res* 21: 939-957.

Chen ZQ, Liu Q, Zhu YS, Li YX (2002) Assessment on predicting methods for membrane-protein transmembrane regions. *Acta Biochem Biophys Sin* 34: 285–290.

Cheng ZQ, Targolli J, Huang XQ, Wu R (2002) Wheat LEA genes, *PMA80* and *PMA1959*, enhance dehydration tolerance of transgenic rice (*Oryza sativa* L.). *Mol Breed* 10: 71–82.

Claes B, Dekeyser R, Villarroel R, Van den Bulcke M, Bauw G, Van Montagu M, et al. (1990) Characterization of a rice gene showing organ-specific expression in response to salt stress and drought. *Plant Cell* 2: 19-27.

Close TJ (1996) Dehydrins: emergence of a biochemical role of a family of plant dehydration proteins. *Plant Physiol* 97: 795–803.

Close TJ (1997) Dehydrins: a commonality in the response of plants in dehydration and low temperature. *Physiol Plant* 100: 291–296.

Close TJ, Kortt AA, Chandler PM (1989) A cDNA-based comparison of dehydration induced proteins (dehydrins) in barley and corn. *Plant Mol Biol* 13: 95-108.

Colmenero-Flores JM, Moreno LP, Smith C, Covarrubias AA (1999) Pvlea-18, a member of a new late-embryogenesis-abundant protein family that accumulates during water stress and in the growing regions of well-irrigated bean seedlings. *Plant Physiol* 120: 93–103.

Colmenero-Flores JM, Campos F, Garciarrubio A, Covarrubias AA (1997) Characterization of *Phaseolus vulgaris* cDNA clones responsive to water deficit: identification of a novel late embryogenesis abundant-like protein. *Plant Mol Biol* 35: 393-405.

Crosatti C, de Laureto PP, Bassi R, Cattivelli L (1999) The interaction between cold and light controls the expression of the cold-regulated barley gene *cor14b* and the accumulation of the corresponding protein. *Plant Physiol* 119: 671-680.

Cuming AC (1999) LEA proteins. In: Casey R, Shewry PR (Eds) Seed proteins. Kluwer Academic Publishers, Dordrecht, The Netherlands, pp 753–780.

Dalal M, Tayal D, Chinnusamy V, Bansal KC (2009) Abiotic stress and ABA-inducible Group 4 LEA from *Brassica napus* plays a key role in salt and drought tolerance. *J Biotechnol* 139: 137-145.

Danyluk J, Perron A, Houde M, Limin A, Fowler B, Benhamou N, et al. (1998) Accumulation of an acidic dehydrin in the vicinity of the plasma membrane during cold acclimation of wheat. *Plant Cell* 10: 623-638.

Delseny M, Bies-Etheve N, Carles C, Hull G, Vicient C, Raynal M, et al. (2001) Late abundant (LEA) protein gene regulation during *Arabidopsis* seed maturation. *J Plant Physiol* 158: 419-427.

Dominguez PG, Frankel N, Mazuch J, Balbo I, Iusem N, Fernie AR, et al. (2013) ASR1 mediates glucose-hormone cross talk by affecting sugar trafficking in tobacco plants. *Plant Physiol* 161: 1486–1500.

Du D, Zhang Q, Cheng T, Pan H, Yang W, Sun L (2013) Genome-wide identification and analysis of late embryogenesis abundant (*LEA*) genes in *Prunus mume*. *Mol Biol Rep* 40: 1937-1946.

Duan J, Cai W (2012) *OsLEA3-2*, an abiotic stress induced gene of rice plays a key role in salt and drought tolerance. *PLOS ONE* 7: 9.

Dure L, 3rd (1993) A repeating 11-mer amino acid motif and plant desiccation. *Plant J* 3: 363-369.

Dure L, III, Crouch M, Harada J, Ho T-H D, Mundy J, Quatrano RS, et al. (1989) Common amino acid sequence domains among the LEA proteins of higher plants. *Plant Mol Biol* 12: 475–486.

Figueras M, Pujal J, Saleh A, Save R, Pages M, Goday A (2004) Maize Rab17 overexpression in *Arabidopsis* plants promotes osmotic stress tolerance. *Ann Appl Biol* 144: 251–257.

Fowler DB, Breton G, Limin AE, Mahfoozi S, Sarhan F (2001) Photoperiod and temperature interactions regulate low temperature-induced gene expression in barley. *Plant Physiol* 127: 1676-1681.

Fu D, Huang B, Xiao Y, Muthukrishnan S, Liang GH (2007) Overexpression of barley *hva1* gene in creeping bentgrass for improving drought tolerance. *Plant Cell Rep* 26: 467-477.

Fujita M, Fujita Y, Maruyama K, Seki M, Hiratsu K, Ohme-Takagi M, et al. (2004) A dehydration-induced NAC protein, RD26, is involved in a novel ABA-dependent stress-signaling pathway. *Plant J* 39: 863-876.

Galau GA, Wang HY-C, Hughes DW (1993) Cotton Lea5 and Lea4 encode atypical late embryogenesis-abundant proteins. *Plant Physiol* 101: 695–696.

Ganguly M, Datta K, Roychoudhury A, Gayen D, Sengupta DN, et al. (2012) Overexpression of *Rab16A* gene in indica rice variety for generating enhanced salt tolerance. *Plant Signal Behav* 7: 502–509.

Ganguly M, Roychoudhury A, Sarkar SN, Sengupta DN, Datta SK, et al. (2011) Inducibility of three salinity/abscisic acid-regulated promoters in transgenic rice with *gusA* reporter gene. *Plant Cell Rep* 30: 1617–1625.

Gao C, Wang C, Zheng L, Wang L, Wang Y (2012) A *LEA* gene regulates cadmium tolerance by mediating physiological responses *Int J Mol Sci* 13: 5468-5481.

Gao W, Bai S, Li Q, Gao C, Liu G, Li G, et al. (2013) Overexpression of *TaLEA* gene from *Tamarix androssowii* improves salt and drought tolerance in transgenic poplar (*Populus simonii* × *P. nigra*). *PLOS ONE* 8: 6.

Garay-Arroyo A, Colmenero-Flores JM, Garciarrubio A, Covarrubias AA (2000) Highly hydrophilic proteins in prokaryotes and eukaryotes are common during conditions of water deficit. *J Biol Chem* 275: 5668–5674.

George S, Usha B, Parida A (2009) Isolation and characterization of an atypical LEA protein coding cDNA and its promoter from drought-tolerant plant *Prosopis juliflora*. *Appl Biochem Biotechnol* 57: 244-253.

Gepts P, Beavis WD, Brummer S, Shoemaker RC, Stalker HT, Weeden NF, et al. (2005) Legumes as a model plant family. Genomics for food and feed report of the cross-legume advances through genomics conference. *Plant Physiol* 137: 1228–1235.

Goday A, Jensen AB, Culianez-Marcia F, Alba MM, Figueras M, Serratosa J, et al. (1994) The maize abscisic acid-responsive protein Rab17 is located in the nucleus and interacts with nuclear localization signals. *Plant Cell* 6: 351-360.

Godoy JA, Pardo JM, Pintor-Toro JA (1990) A tomato cDNA inducible by salt stress and abscisic acid: nucleotide sequence and expression pattern. *Plant Mol Biol* 15: 695-705.

Gomez J, Sanchez-Martinez D, Stiefel V, Rigau J, Puigdomenech P, Pages M (1988) A gene induced by the plant hormone abscisic acid in response to water stress encodes a glycine-rich protein. *Nature* 334: 262-264.

Goyal K, Pinelli C, Maslen SL, Rastogi RK, Stephens E, Tunnacliffe A (2005a) Dehydration-regulated processing of late embryogenesis abundant protein in a desiccation-tolerant nematode. *FEBS Lett* 579: 4093–4098.

Goyal K, Walton LJ, Tunnacliffe A (2005b) LEA proteins prevent protein aggregation due to water stress. *Biochem J* 388: 151–157.

Grelet J, Benamar A, Teyssier E, Avelange-Macherel MH, Grunwald D, Macherel D (2005) Identification in pea seed mitochondria of a late-embryogenesis abundant protein able to protect enzymes from drying. *Plant Physiol* 137: 157–167.

Gu H, Jia Y, Wang X, Chen Q, Shi S, Ma L, et al. (2012) Identification and characterization of a LEA family gene *CarLEA4* from chickpea (*Cicer arietinum* L.). *Mol Biol Rep* 39: 3565-3572.

Hara M (2010) The multifunctionality of dehydrins. An overview. *Plant Signal Behav* 5: 503–508.

Hara M, Fujinaga M, Kuboi T (2004) Radical scavenging activity and oxidative modification of citrus dehydrin. *Plant Physiol Biochem* 42: 657-662.

Hara M, Fujinaga M, Kuboi T (2005) Metal binding by citrus dehydrin with histidine-rich domains. *J Exp Bot* 56: 2695–2703.

Hara M, Shinoda Y, Tanaka Y, Kuboi T (2009) DNA binding of citrus dehydrin promoted by zinc ion. *Plant Cell Environ* 32: 532–541.

Hara M, Terashima S, Fukaya T, Kuboi T (2003) Enhancement of cold tolerance and inhibition of lipid peroxidation by citrus dehydrin in transgenic tobacco. *Planta* 203: 290-298.

Hara M, Terashima S, Kuboi T (2001) Characterization and cryoprotective activity of cold-responsive dehydrin from *Citrus unshiu*. *J Plant Physiol* 158: 1333-1339.

Hara M, Wakasugi Y, Ikoma Y, Yano M, Ogawa K, Kuboi T (1999) cDNA sequence and expression of a cold-responsive gene in *Citrus unshiu*. *Biosci Biotechnol Biochem* 63: 433–437.

Harada JJ, Delisle AJ, Baden CS, Crouch ML (1989) 30 Unusual sequence of an abscisic acid-inducible mRNA which accumulates late in *Brassica napus* seed development *Plant Mol Biol* 12: 395-401.

Hattori T, Terada T, Hamasuna S (1995) Regulation of *Osem* gene and the transcriptional activator VP1: analysis of c*is*-acting promoter elements required for regulation by abscisic acid and VP1. *Plant J* 7: 913-925.

He S, Tan L, Hu Z, Chen G, Wang G, Hu T (2012) Molecular characterization and functional analysis by heterologous expression in *E. coli* under diverse abiotic stresses for OsLEA5, the atypical hydrophobic LEA protein from *Oryza sativa* L. *Mol Genet Genomics* 287: 39-54.

Herzer S, Kinealy K, Asbury R, Beckett P, Eriksson K, Moore P (2003) Purification of native dehydrin from *Glycine max* cv., *Pisum sativum* and *Rosmarinum officinalis* by affinity chromatography. *Protein Expr Purif* 28: 232–240.

Hinniger C, Caillet V, Michoux F, Amor MB, Tanksley S, Lin C, et al. (2006) Isolation and characterization of cDNA encoding three dehydrins expressed during *Coffea canephora* (robusta) grain development. *Ann Bot* 97: 755–765.

Hobo T, Asada M, Kowyama Y, Hattori T (1999) ACGT-containing abscisic acid response element (ABRE) and coupling element 3 (CE3) are functionally equivalent. *Plant J* 19: 679–689.

Hong B, Uknes SJ, Ho T-HD (1988) Cloning and characterization of a cDNA encoding a mRNA rapidly induced by ABA in barley aleurone layers. *Plant Mol Biol* 11: 495-506.

Hong-Bo S, Zong-Suo L, Ming-An S (2005) LEA proteins in higher plants: structure, function, gene expression and regulation. Colloids and Surfaces B: *Biointerfaces* 45: 131–135.

Horvath DP, McLarney BK, Thomashow MF (1993) Regulation of *Arabidopsis thaliana* L. (Heyn) *cor78* in response to low temperature. *Plant Physiol* 103: 1047-1053.

Houde M, Dallaire S, N'Dong D, Sarhan F (2004) Overexpression of the acidic dehydrin WCOR410 improves freezing tolerance in transgenic strawberry leaves. *Plant Biotech J* 2: 381–387.

Houde M, Daniel C, Lachapelle M, Allard F, Laliberte S, Sarhan F (1995) Immunolocalization of freezing-tolerance-associated proteins in the cytoplasm and nucleoplasm of wheat crown tissues. *Plant J* 8: 583–593.

Hsing YIC, Tsou CH, Hsu TF, Chen ZY, Hsieh KL, Hsieh JS, et al. (1998) Tissue- and stage-specific expression of a soybean (*Glycine max* L.) seed-maturation, biotinylated protein. *Plant Mol Biol* 38: 481–490.

Hundertmark M, Hincha DK (2008) LEA (Late Embryogenesis Abundant) proteins and their encoding genes in *Arabidopsis thaliana. BMC Genomics* 9: 118.

Ingram J, Bartels D (1996) The molecular basis of dehydration tolerance in plants. *Annu Rev Plant Physiol Plant Mol Biol* 47: 377–403.

Ismail AM, Hall AE, Close TJ (1997) Chilling tolerance during emergence of cowpea associated with a dehydrin and slow electrolyte leakage. *Crop Sci* 37: 1270–1277.

Ismail AM, Hall AE, Close TJ (1999a) Purification and partial characterization of a dehydrin involved in chilling tolerance during seedling emergence of cowpea. *Plant Physiol* 120: 237-244.

Ismail AM, Hall AE, Close TJ (1999b) Allelic variation of a dehydrin gene cosegregates with chilling tolerance during seedling emergence. *Proc Natl Acad Sci USA* 96: 13566-13570.

Iwasaki T, Kiyosue T, Yamaguchi-Shinozaki K, Shinozaki K (1997) The dehydration-inducible *Rd17* (*Cor47*) gene and its promoter region in *Arabidopsis thaliana* (Accession No. AB004872). Plant Gene Register. *Plant Physiol* 115: 1287-1289.

Jensen AB, Goday A, Figueras M, Jessop AC, Pages M (1998) Phosphorylation mediates the nuclear targeting of the maize Rab17 protein. *Plant J* 13: 691-697.

Kim H-S, Lee JH, Kim JJ, Kim C-H, Jun S-S, Hong Y-N (2005) Molecular and functional characterization of *CaLEA6*, the gene for a hydrophobic LEA protein from *Capsicum annuum*. *Gene* 344: 115–123.

King SW, Joshi CP, Naguyen HT (1992) DNA sequence of an ABA-responsive gene (*rab15*) from water stressed wheat roots. *Plant Mol Biol* 18: 377-383.

Kirch H-H, Van Berkel J, Glaczinski H, Salamini F, Gebhardt C (1997) Structural organization, expression and promoter activity of a cold-stress-inducible gene of potato (*Solanum tuberosum* L.). *Plant Mol Biol* 33: 897-909.

Kiyosue T, Yamaguchi-Shinozaki K, Shinozaki K (1994) Characterization of two cDNAs (*ERD10* and *ERD14*) corresponding to genes that respond rapidly to dehydration stress in *Arabidopsis thaliana*. *Plant Cell Physiol* 35: 225-231.

Kizis D, Pages M (2002) Maize DRE-binding proteins DBF1 and DBF2 are involved in *rab17* regulation through the drought-responsive element in an ABA-dependent pathway. *Plant J* 30: 679–689.

Ko S, Kamada H (2002) Isolation of carrot basic leucine zipper transcription factor using yeast one-hybrid screening. *Plant Mol Biol Rep* 20: 301a-301h.

Koag MC, Fenton RD, Wilkens S, Close TJ (2003) The binding of maize DHN1 to lipid vesicles. Gain of structure and lipid specificity. *Plant Physiol* 131: 309.316.

Koag MC, Wilkens S, Fenton RD, Resnik J, Vo E, Close TJ (2009) The K-segment of maize DHN1 mediates binding to anionic phospholipid vesicles and concomitant structural changes. *Plant Physiol* 150: 1503–1514.

Koike M, Takezawa D, Arakawa K, Yoshida S (1997) Accumulation of 19-kDa plasma membrane polypeptide during induction of freezing tolerance in wheat suspension-cultured cells by abscisic acid. *Plant Cell Physiol* 38: 707–716.

Koizumi M, Yamaguchi-Shinozaki K, Tsuji H, Shinozaki K (1993) Structure and expression of two genes that encode distinct drought-inducible cysteine proteinases in *Arabidopsis thaliana*. *Gene* 129: 175-182.

Kontunen-Soppela S, Taulavuori K, Taulavuori E, Lähdesmäki P, Laine K (2000) Soluble proteins and dehydrins in nitrogen-fertilized Scots pine seedlings during deacclimation and the onset of growth. *Physiol Plant* 109: 404-409.

Kosová K, Vítámvás P, Prášil IT (2007) The role of dehydrins in plant response to cold. *Biol Plant* 51: 601–617.

Kruger C, Berkowitz O, Stephan UW, Hell R (2002) A metal-binding member of the late embryogenesis abundant protein family transports iron in the phloem of *Ricinus communis* L. *J Biol Chem* 277: 25062–25069.

Kusano T, Aguan K, Abe M, Sugawara K (1992) Nucleotide sequence of a rice *rab16* homologue gene. *Plant Mol Biol* 18: 127-129.

Kwapata K, Nguyen T, Sticklen M (2012) Genetic transformation of common bean (*Phaseolus vulgaris* L.) with the gus color marker, the bar herbicide resistance, and the barley (*Hordeum vulgare*) *HVA1* drought tolerance genes. *Int J Agron Article* ID 198960, 8 pages.

Lal S, Gulyani V, Khurana P (2008) Overexpression of *HVA1* gene from barley generates tolerance to salinity and water stress in transgenic mulberry (*Morus indica*). *Transgenic Res* 17: 651-663.

Lang V, Palva ET (1992) The expression of a *rab*-related gene, *rab18*, is induced by abscisic acid during the cold acclimation process of *Arabidopsis thaliana* (L.) Heynh. *Plant Mol Biol* 20: 951-962.

Lee S-C, Lee M-Y, Kim S-J, Jun S-H, An G, Kim S-R (2005) Characterization of an abiotic stress-inducible dehydrin gene, *OsDhn1*, in rice (*Oryza sativa* L.). *Mol Cells* 19: 212-218.

Li NY, Gao JF, Wang PH (1998) Protein characters induced by water stresses in wheat seedlings. *Acta Phytophysiol Sin* 24: 65–71.

Lin CH, Peng PH, Ko CY, Markhart AH, Lin TY (2012) Characterization of a novel $Y_2$K-type dehydrin VrDhn1 from *Vigna radiata*. *Plant Cell Physiol* 53: 930–942.

Liu G, Xu H, Zhang L, Zheng Y (2011) Fe binding properties of two soybean (*Glycine max* L.) LEA4 proteins associated with antioxidant activity. *Plant Cell Physiol* 52: 994–1002.

Liu Y, Wang L, Sun L, Pan J, Kong X, Zhang M, et al. (2013) ZmLEA3, a multifunctional group 3 LEA protein from maize *Zea mays* L. is involved in biotic and abiotic stresses. *Plant Cell Physiol* 54: 944–959.

Liu Y, Zheng Y, Zhang Y, Wang W, Li R (2010) Soybean PM2 protein (LEA3) confers the tolerance of *Escherichia coli* and stabilization of enzyme activity under diverse stresses. *Curr Microbiol* 60: 373–378.

Manfre AJ, Lahatte GA, Climer CR, Marcotte WR, Jr (2009) Seed dehydration and the establishment of desiccation tolerance during seed maturation is altered in the *Arabidopsis thaliana* mutant *atem6-1*. *Plant Cell Physiol* 50: 243–253.

Manfre AJ, Lanni LM, Marcotte WR, Jr (2006) The *Arabidopsis* group 1 LATE EMBRYOGENESIS ABUNDANT protein ATEM6 is required for normal seed development. *Plant Physiol* 140: 140 –149.

Marcotte WR, Jr, Bayley CC, Quatrano RS (1988) Regulation of a wheat promoter by abscisic acid in rice protoplasts. *Nature* 335: 454-457.

Maskin L, Frankel N, Gudesblat G, Demergasso MJ, Pietrasanta LI, Iusem ND (2007). Dimerization and DNA-binding of ASR1, a small hydrophilic protein abundant in plant tissues suffering from water loss. *Biochem Biophys Res Com* 352: 831–835.

Miller JD, Arteca RN, Pell EJ (1999). Senescence-associated gene expression during ozone-induced leaf senescence in *Arabidopsis. Plant Physiol* 120: 1015–1024.

Momma M, Kaneko S, Haraguchi K, Matsukura U (2003) Peptide mapping and assessment of cryoprotective activity of 26/27-kDa dehydrin from soybean seeds. *Biosci Biotechnol Biochem* 67: 1832–1835.

Monroy AF, Castonguay Y, Laberge S, Sarhan F, Vezina LP, Dhindsa RS (1993) A new cold-induced alfalfa gene is associated with enhanced hardening at subzero temperature. *Plant Physiol* 102: 873-879.

Moons A, Bauw G, Prinsen E, Van Montagu M, Van der Straeten D (1995) Molecular and physiological responses to abscisic acid and salts in roots of salt-sensitive and salt-tolerant indica rice varieties. *Plant Physiol* 107: 177-186.

Moons A, De Keyser A, Van Montagu M (1997) A group 3 LEA cDNA of rice, responsive to abscisic acid, but not to jasmonic acid, shows variety-specific differences in salt stress response. *Gene* 191: 197-204.

Moreno-Fonseca LP, Covarrubias AA (2001) Downstream DNA sequences are required to modulate *Pvlea-18* gene expression in response to dehydration. *Plant Mol Biol* 45: 501–515.

Morris PC, Kumar A, Bowles DJ, Cuming AC (1990) Osmotic stress and abscisic acid induce expression of the wheat *Em* genes. *Eur J Biochem* 190: 625-630.

Mowla SB, Cuypers A, Driscoll SP, Kiddle G, Thomson J, Foyer CH, et al. (2006) Yeast complementation reveals a role for an *Arabidopsis thaliana* late embryogenesis abundant (LEA)-like protein in oxidative stress tolerance. *Plant J* 48: 743–756.

Mukherjee K, Roy Choudhury A, Gupta B, Gupta S, Sengupta DN (2006) An ABRE-binding factor, OSBZ8, is highly expressed in salt tolerant cultivars than in salt sensitive cultivars of indica rice. *BMC Plant Biol* 6:18.

Mundy J, Chua NH (1988) Abscisic acid and water-stress induce the expression of a novel rice gene. *EMBO J* 7: 2275-2287.

Muñoz-Mayor A, Pineda B, Garcia-Abellán JO, Antón T, Garcia-Sogo B, Sanchez-Bel P, et al. (2012) Overexpression of dehydrin *tas14* gene improves the osmotic stress imposed by drought and salinity in tomato. *J Plant Physiol* 169: 459-468.

Nakashima K, Ito Y, Yamaguchi-Shinozaki K (2009) Transcriptional regulatory networks in response to abiotic stresses in *Arabidopsis* and grasses. *Plant Physiol* 149: 88–95.

Naot D, Ben-Hayyim G, Eshdat Y, Holland D (1995) Drought, heat, and salt stress induce the expression of a citrus homologue of an atypical late embryogenesis *Lea5* gene. *Plant Mol Biol* 27: 619–622.

N'Dong C, Danyluk J, Wilson KE, Pocock T, Huner NPA, Sarhan F (2002) Cold-regulated cereal chloroplast late embryogenesis abundant-like proteins. Molecular characterization and functional analyses. *Plant Physiol* 129: 1368-1381.

Neven LG, Haskell DW, Hofig A, Li QB, Guy CL (1993) Characterization of a spinach gene responsive to low temperature and water stress. *Plant Mol Biol* 21: 291-305.

Niogret MF, Culianez-Macia FA, Goday A, Alba MM, Pages M (1996) Expression and cellular localization of *rab28* mRNA and Rab28 protein during maize embryogenesis. *Plant J* 9: 549-557.

Olvera-Carrillo Y, Reyes JL, Covarrubias AA (2011) Late embryogenesis abundant proteins: versatile players in the plant adaptation to water limiting environments. *Plant Signal Behav* 64: 1–4.

Park BJ, Liu Z, Kanno A, Kameya T (2005) Transformation of radish (*Raphanus sativus* L.) via sonication and vacuum infiltration of germinated seeds with *Agrobacterium* harboring a group 3 *LEA* gene from *B. napus*. *Plant Cell Rep* 24: 494-500.

Plant AL, Cohen A, Moses MS, Bray EA (1991) Nucleotide sequence and spatial expression pattern of a drought and abscisic acid-induced gene of tomato. *Plant Physiol* 97: 900-906.

Porat R, Pasentsis K, Rozentzvieg D, Gerasopoulos D, Falara V, Samach A, et al. (2004) Isolation of a dehydrin cDNA from orange and grapefruit citrus fruit that is specifically induced by the combination of heat followed by chilling temperatures. *Physiol Plant* 120: 256-264.

Porat R, Pavoncello D, Lurie S, McCollum TG (2002) Identification of a grapefruit cDNA belonging to a unique class of citrus dehydrins and characterization of its expression patterns under temperature stress conditions. *Physiol Plant* 115: 598-603.

Puhakainen T, Hess MV, Makela P, Svensson J, Heino P, Palva ET (2004) Overexpression of multiple dehydrin genes enhances tolerance to freezing stress in *Arabidopsis*. *Plant Mol Biol* 54: 743-753.

Raynal M, Guilleminot J, Gueguen C, Cooke R, Delseny M, Gruber V (1999) Structure, organization and expression of two closely related novel Lea (late-embryogenesis-abundant) genes in *Arabidopsis thaliana*. *Plant Mol Biol* 40: 153–165.

Reddy PS, Reddy GM, Pandey P, Chandrasekhar K, Reddy MK (2012) Cloning and molecular characterization of a gene encoding late embryogenesis abundant protein from *Pennisetum glaucum*: protection against abiotic stresses. *Mol Biol Rep* 39: 7163-7174.

Reyes JL, Campos F, Wei H, Arora R, Yang Y, Karlson DT, et al. (2008) Functional dissection of hydrophilins during *in vitro* freeze protection. *Plant Cell Environ* 31: 1781–1790.

Reyes JL, Rodrigo MJ, Colmenero-Flores JM, Gil JV, Garay-Arroyo A, Campos F, et al. (2005) Hydrophilins from distant organisms can protect enzymatic activities from water limitation effects *in vitro*. *Plant Cell Environ* 28: 709–718.

Richard S, Morency M-J, Drevet C, Jouanin L, Seguin A (2000) Isolation and characterization of a dehydrin gene from white spruce induced upon wounding, drought and cold stresses. *Plant Mol Biol* 43: 1-10.

Rinne PLH, Kaikuranta PLM, van der Plas LHW, van der Schoot C (1999) Dehydrins in cold-acclimated apices of birch (*Betula pubescens* Ehrh.): production, localization and potential role in rescuing enzyme function during dehydration. *Planta* 209: 377–388.

Rodriguez EM, Svensson JT, Malatrasi M, Choi D-W, Close TJ (2005) Barley *Dhn13* encodes a KS-type dehydrin with constitutive and stress responsive expression. *Theor Appl Genet* 110: 852-858.

Rorat T (2006) Plant dehydrins - Tissue localization, structure and function. *Cell Mol Biol Lett* 11: 536–556.

Rorat T, Grygorowicz WJ, Irzykowski W, Rey P (2004) Expression of KS-type dehydrins is primarily regulated by factor related to organ type and leaf developmental stage during vegetative growth. *Planta* 218: 878-885.

Rorat T, Szabala BM, Grygorowicz WJ, Wojtowicz B, Yin Z, Rey P (2006) Expression of SK$_3$-type dehydrin in transporting organs is associated with cold acclimation in *Solanum* species. *Planta* 224: 205-221.

Roychoudhury A, Basu S, Sengupta DN (2009) Comparative expression of two abscisic acid-inducible genes and proteins in seeds of aromatic indica rice cultivar with that of non-aromatic indica rice cultivars. *Indian J Exp Biol* 47: 827-833.

Roychoudhury A, Gupta B, Sengupta DN (2008) Trans-acting factor designated OSBZ8 interacts with both typical abscisic acid responsive elements as well as abscisic acid responsive element-like sequences in the vegetative tissues of indica rice cultivars. *Plant Cell Rep* 27: 779-794.

Roychoudhury A, Paul A (2012) Abscisic acid-inducible genes during salinity and drought stress. In: Berhardt LV (eds) Advances in medicine and biology, vol 51. Nova Publishers, New York, pp 1–78.

Roychoudhury A, Paul S, Basu S (2013) Cross-talk between abscisic acid-dependent and abscisic acid-independent pathways during abiotic stress. *Plant Cell Rep* 32: 985–1006.

RoyChoudhury A, Roy C, Sengupta DN (2007) Transgenic tobacco plants overexpressing the heterologous *lea* gene *Rab16A* from rice during high salt and water deficit display enhanced tolerance to salinity stress. *Plant Cell Rep* 26: 1839–1859.

Roychoudhury A, Sengupta DN (2009) The promoter-elements of some abiotic stress-inducible genes from cereals interact with a nuclear protein from tobacco. *Biol Plant* 53: 583-587.

Sarhan F, Ouellet F, Vazquez-Tello A (1997) The wheat *wcs120* gene family. A useful model to understand the molecular genetics of freezing tolerance in cereals. *Physiol Plant* 101: 439-445.

Sarnighausen E, Karlson D, Aahworth E (2002) Seasonal regulation of a 24-kDa protein from red-osier dogwood (*Cornus sericea*) xylem. *Tree Physiol* 22: 423-430.

Savitri ES, Basuki N, Aini N, Arumingtyas EL (2013) Identification and characterization drought tolerance of gene *LEA-D11* soybean (*Glycine max* L. Merr) based on PCR-sequencing. *Am J Mol Biol* 3: 32-37.

Sharon MA, Kozarova A, Clegg JS, Vacratsis PO, Warner AH (2009). Characterization of a group 1 late embryogenesis abundant protein in encysted embryos of the brine shrimp *Artemia franciscana*. *Biochem Cell Biol* 87: 415-430.

Shen Q, Chen C-N, Brands A, Pan S-M, Ho T-HD (2001) The stress and abscisic acid-induced barley gene *HVA22*: developmental regulation and homologues in diverse organisms. *Plant Mol Biol* 45: 327-340.

Shen Q, Uknes SJ, Ho T-HD (1993) Hormone response complex in a novel abscisic acid and cycloheximide-inducible barley gene. J Biol Chem 268: 23652–23660.

Shi M-D, Hoekstra FA, Hsing Y-IC (2008) Late Embryogenesis Abundant Proteins. *Adv Bot Res* 48: 211-255.

Shih M-D, Huang L-T, Wei F-J, Wu M-T, Hoekstra FA, Hsing Y-IC (2010) OsLEA1a, a new Em-like protein of cereal plants. *Plant Cell Physiol* 51: 2132–2144.

Shih MD, Lin SC, Hsieh JS, Tsou CH, Chow TY, Lin TP, et al. (2004) Gene cloning and characterization of a soybean (*Glycine max* L.) LEA protein, GmPM16. *Plant Mol Biol* 56: 689–703.

Shinde S, Shinde R, Downey F, Ng C (2013) Abiotic stress-induced oscillations in steady-state transcript levels of Group 3 *LEA* protein genes in the moss, *Physcomitrella patens Plant Signal Behav* 8:1 e22535-78-e22535-82.

Singh S, Cornilescu CC, Tyler RC, Cornilescu G, Tonelli M, Lee MS, et al. (2005) Solution structure of a late embryogenesis abundant protein (LEA14) from *Arabidopsis thaliana*, a cellular stress-related protein. *Protein Sci* 14: 2601–2609.

Sivamani E, Bahieldin A, Wraith JM, Al-Niemi T, Dyer WE, Ho THD, et al. (2000) Improved biomass productivity and water use efficiency under water deficit conditions in transgenic wheat constitutively expressing the barley *HVA1* gene. *Plant Sci* 155: 1–9.

Skriver K, Mundy J (1990) Gene expression in response to abscisic acid and osmotic stress. *Plant Cell* 2: 503-512.

Solomon A, Salomon R, Paperna I, Glazer I (2000) Desiccation stress of entomopathogenic nematodes induces the accumulation of a novel heat-stable protein. *Parasitology* 121: 409-416.

Stacy RA, Nordeng TW, Culianez-Macia FA, Aalen RB (1999) The dormancy-related peroxiredoxin anti-oxidant, PER1, is localized to the nucleus of barley embryo and aleurone cells. *Plant J* 19: 1-8.

Straub PF, Shen Q, Ho THD (1994) Structure and promoter analysis of an ABA- and stress-regulated barley gene *HVA1*. *Plant Mol Biol* 26: 617-630.

Su L, Zhao C-Z, Bi Y-P, Wan S-B, Wang X-J (2004) Isolation and expression analysis of *LEA* genes in peanut (*Arachis hypogaea* L.). *Plant Mol Biol* 54: 743–753.

Svensson J, Palva ET, Welin B (2000) Purification of recombinant *Arabidopsis thaliana* dehydrins by metal ion affinity chromatography. *Protein Expr Purif* 20: 169–178.

Swamy PM, Smith BN (1999) Role of abscisic acid in plant stress tolerance. *Curr Sci* 76: 1220-1227.

Takahashi R, Joshee N, Kitagawa Y (1994) Induction of chilling resistance by water stress and cDNA sequence analysis and expression of water stress regulated genes in rice. *Plant Mol Biol* 26: 339–352.

Takumi S, Koike A, Nakata M, Kume S, Ohno R, Nakamura C (2003) Cold-specific and light-stimulated expression of a wheat (*Triticum aestivum* L.) *Cor* gene *Wcor15* encoding a chloroplast-targeted protein. *J Exp Bot* 54: 2265-2274.

Tanaka S, Ikeda K, Miyasaka H (2004) Isolation of a new member of group 3 late embryogenesis abundant protein gene from a halotolerant green alga by a functional expression screening with cyanobacterial cells. *FEMS Microbiol Lett* 236: 41-45.

Thomashow MF (1994) *Arabidopsis thaliana* as a model for studying mechanisms of plant cold tolerance. In: Meyerowitz E, Somerville C (eds) *Arabidopsis*, Cold Spring Harbor, New York: Cold Spring Harbor Laboratory Press, 807-834.

Thomashow MF (1998) Role of cold-responsive genes in plant freezing tolerance. *Plant Physiol* 118: 1-8.

Tsvetanov S, Ohno R, Tsuda K, Takumi S, Mori N, Atanassov A, et al. (2000) A cold-responsive wheat (*Triticum aestivum* L.) gene *wcor14* identified in a winter-hardy cultivar 'Mironovska 808'. *Genes Genet Syst* 75: 49-57.

Tunnacliffe A, Wise MJ (2007) The continuing conundrum of the LEA proteins. *Naturwissenschaften* 94: 791-812.

Vernon DM, Ostrem JA, Bohnert HJ (1993) Stress perception and response in a facultative halophyte – The regulation of salinity-induced genes in *Mesembryanthemum crystallinum*. *Plant Cell Environ* 16: 437-444.

Villardell J, Goday A, Freire MA, Torrent M, Martinez MC, Torne JM, Pages M (1990) Gene sequence, developmental expression, and protein phosphorylation of Rab17 in maize. *Plant Mol Biol* 14: 423-432.

Wang L, Li X, Chen S, Liu G (2009) Enhanced drought tolerance in transgenic *Leymus chinensis* plants with constitutively expressed wheat TaLEA3. *Biotechnol Lett* 31: 313-319.

Wang X-S, Zhu H-B, Jin G-L, Liu H-L, Wu W-R, Zhu J (2007) Genome-scale identification and analysis of *LEA* genes in rice (*Oryza sativa* L.). *Plant Sci* 172: 414–420.

Wellin BV, Olson A, Nylander M, Palva ET (1994) Characterization and differential expression of *Dhn/Lea/Rab*-like genes during cold acclimation and drought stress in *Arabidopsis thaliana*. *Plant Mol Biol* 26: 131-144.

Wellin BV, Olson A, Palva ET (1995) Structure and organization of two closely-related low-temperature-induced *Dhn/Lea/Rab*-like genes in *Arabidopsis thaliana* (L.) *Heynh*. *Plant Mol Biol* 29: 391-395.

Whitsitt MS, Collins RG, Mullet JE (1997) Modulation of dehydration tolerance in soybean seedlings (Dehydrin Mat1 is induced by dehydration but not by abscisic acid). *Plant Physiol* 114: 917–925.

Wise MJ (2003) LEAping to conclusions: A computational reanalysis of late embryogenesis abundant proteins and their possible roles. *BMC Bioinformatics* 4: 52–70.

Wise MJ, Tunnaclife A (2004) POPP the question: what do LEA proteins do? *Trends Plant* Sci 9: 13-17.

Wolfraim LA, Langis R, Tyson H, Dhindsa RS (1993) cDNA sequence, expression, and transcript stability of a cold acclimation-specific gene, *cas18*, of alfalfa (*Medicago falcata*) cells. *Plant Physiol* 101: 1275-1282.

Wolkers WF, McCready S, Brandt WF, Lindsey GG, Hoekstra FA (2001) Isolation and characterization of a D-7 LEA protein from pollen that stabilizes glasses in vitro. *Biochim Biophys Acta* 1544: 196–206.

Xiao B, Huang Y, Tang N, Xiong L (2007) Over-expression of a *LEA* gene in rice improves drought resistance under the field conditions. *Theor Appl Genet* 115: 35–46.

Xing X, Liu Y, Kong X, Liu Y, Li D (2012) Overexpression of a maize dehydrin gene, *ZmDHN2b*, in tobacco enhances tolerance to low temperature. *Plant Growth Regul* 65: 109-118.

Xu D, Duan X, Wang B, Hong B, Ho THD, Ray W (1996) Expression of a late embryogenesis abundant protein gene, *HVA1* from barley confers tolerance to water deficit and salt stress in transgenic rice. *Plant Physiol* 110: 249–257.

Yamaguchi-Shinozaki K, Mundy J, Chua N-H (1989) Four tightly linked *rab* genes are differentially expressed in rice. *Plant Mol Biol* 14: 29-39.

Yamaguchi-Shinozaki K, Shinozaki K (1993a) The plant hormone abscisic acid mediates the drought-induced expression but not the seed-specific expression of *rd22,* a gene responsive to dehydration stress in *Arabidopsis thaliana*. *Mol Gen Genet* 238: 17-25.

Yamaguchi-Shinozaki K, Shinozaki K (1993b) Characterization of the expression of a desiccation-responsive *rd29* gene of *Arabidopsis thaliana* and analysis of its promoter in transgenic plants. *Mol Gen Genet* 236: 331–340.

Yamaguchi-Shinozaki K, Shinozaki K (1994) A novel cis-acting element in an *Arabidopsis* gene is involved in responsiveness to drought, low-temperature, or high-salt stress. *Plant Cell* 6: 251–264.

Yang L, Li W, Xin X, Liping S, Jiaowen P, Xiangpei K, et al. (2013) ZmLEA3, a multifunctional group 3 LEA protein from maize (*Zea mays* L.), is involved in biotic and abiotic stresses. *Plant Cell Physiol* 54: 944-959.

Yao K, Lockhart KM, Kalanack JJ (2005) Cloning of dehydrin sequences from *Brassica juncea* and *Brassica napus* and their low temperature-inducible expression in germinating seeds. *Plant Physiol Biochem* 43: 83-89.

Yin Z, Rorat T, Szabala BM, Ziolkowska A, Malepszy S (2006) Expression of a *Solanum sogarandinum* $SK_3$-type dehydrin enhances cold tolerance in transgenic cucumber seedlings. *Plant Sci* 170: 1164-1172.

Yu JN, Zhang LS, Gao JF (2002) Effect of water stress on induced protein expression of wheat seedlings and suspension cultures, *Acta Bot Boreal-Occident Sin* 52: 865–870.

Zhang LS, Zhao WM (2003) LEA protein functions to tolerance drought of the plant. *Plant Physiol Commun* 39: 61–66.

Zhang Y, Li Y, Lai J, Zhang H, Liu Y, Liang L, et al. (2012) Ectopic expression of a LEA protein gene *TsLEA1* from *Thellungiella salsuginea* confers salt-tolerance in yeast and *Arabidopsis*. *Mol Biol Rep* 39: 4627-4633.

Zhao KY, Yun J, Shi F, Wang J, Yang Q, Chao Y (2010) Molecular cloning and characterization of a group 3 LEA gene from *Agropyron mongolicum*. *Afr J Biotechnol* 9: 6040-6048.

Zhao P, Liu F, Ma M, Gong J, Wang Q, Jia P, et al. (2011) Overexpression of AtLEA3-3 confers resistance to cold stress in *Escherichia coli* and provides enhanced osmotic stress tolerance and ABA sensitivity in *Arabidopsis thaliana. Mol Biol* (Mosk) 45: 851-862.

Zhu B, Choi D-W, Fenton R, Close TJ (2000) Expression of the barley dehydrin multigene family and the development of freezing tolerance. *Mol Gen Genet* 264: 145-153.

In: Abiotic Stress

ISBN: 978-1-63117-622-7

Editor: Annabella Ferro

© 2014 Nova Science Publishers, Inc.

*Chapter 3*

# DEHYDRATION RESPONSIVE ELEMENT BINDING (DREB) TRANSCRIPTION FACTORS AND PLANT ABIOTIC STRESS TOLERANCE

*Donia Bouaziz*[*][1] *and Radhia Gargouri-Bouzid*[1]

[1]Laboratoire des Biotechnologies Végétales Appliquées à l'Amélioration des Cultures, Ecole Nationale d'Ingénieurs de Sfax, BP, Tunisia

## ABSTRACT

Abiotic stresses, such as drought, high salinity, extreme temperatures, heavy metals and oxidative stress affect plant growth and extremely decrease crop productivity.

Plants respond to these environmental challenges through the activation and the regulation of specific stress related genes. The dehydration responsive element binding (DREB) transcription factors, that strongly up-regulates many downstream genes, play an important role in plant environmental stress tolerance, thereby increase efficiency of plant production.

Recently, several research approaches were developed to understand the molecular mechanisms of stress responses based on the investigation of the involvement of such transcription factors on signaling pathways related to abiotic stress tolerance.

The present review discusses the functions of the DREB transcription factors in plant abiotic stress responses and summarizes the recent

---

[*] Corresponding author: donia.bouaziz@yahoo.fr. Fax: +216.74665190.

advances in elucidating stress-response mechanisms and their biotechnological applications. This review examines also the progress of the genetic engineering approaches developed with these DREB transcription factors in the main crops and model plants in order to elucidate their impact on plant tolerance.

**Keywords:** DREB, abiotic stress, stress related genes, Transcription factor

# INTRODUCTION

Abiotic stress conditions such as drought, high salinity and extreme temperatures are constantly threatening plant growth and have adverse effects on biomass production and crop productivity worldwide.

These abiotic stresses lead to series of morphological, physiological, biochemical and molecular changes in plants [1]. The ongoing elucidation of the molecular control mechanisms involved in abiotic stress tolerance, may result in the use of molecular tools for engineering more tolerant plants, by the expression of specific stress-related genes.

Plants developed different pathways in order to overcome the abiotic stresses. They are based on the activation of a variety of signaling pathways starting from activation of secondary messengers, hormones, proteins kinases to the activation of transcription factors that finally allowed expression of responsive genes. Transcription factors play central roles in gene expression by regulating the expression of downstream genes as trans-acting elements via specific binding sequence in the promoters of target genes [2].

Dehydration-responsive element-binding protein (DREB) subfamily belongs to the APETALA 2/ethylene-responsive element binding factor (AP2/ERF) family. The AP2/ERF family is a large group of transcription factors containing AP2/ERF-type DNA binding domain. This latter was first found in the *Arabidopsis* homeotic gene APETALA 2, which is involved in flower Development [3]. Similar domain was then found in tobacco (*Nicotiana tabacum*) ethylene-responsive element binding proteins (EREBPs) [4].

The AP2/ERF family comprises two subfamilies, the EREBP (single AP2 domain) and AP2/family (two copies of AP2). The EREBP subfamily is further divided into two classes, i.e., ERFs and DREBs/CBFs.

The ERF protein binds specifically to Ethylene Responsive Elements (ERE) in the promoter region of the target genes. These ERE elements contain a conserved region known as the GCC box [5].

The ERF proteins possess conserved alanine and aspartic acid residues in the AP2/ERF domain at the $14^{th}$ and $19^{th}$ position respectively that are involved in the DNA binding site recognition.

The DREBs (Drought Responsive Element Binding) proteins harbor a Val-14 and Glu-19, which are important amino acids involved in the binding of DREBs to DRE sequences (TACCGACAT). These sequences are found in the promoters of cold and dehydration responsive genes including LEA (Late Embryogenesis Abundant), *RD29A, RD17* (Responsive to Dehydration), *COR6.6, COR15a* (Cold Responsive), *ERD10* (Early Responsive to Dehydration) and *KIN1* (cold inducible) [2, 6].

Many stress-inducible DREB subfamily members have been isolated and characterized (Table 1). It has been established that they are major factors involved in abiotic stress responses by regulating gene expression via the cis-acting dehydration-responsive element/C-repeat (DRE/CRT) element [2]. However, a number of studies reported that some DREB factors can bind also to GCC box present in the promoter sequence of biotic stress responsive genes [4]. Indeed, study of four AP2/ERFs of *Brassica napus* (BnDREBIII) showed that BnDREBIII 1-3 were able to bind to both GCC box and DRE element while BnDREBIII-4 was not [7].

The first isolated DREB family member was CRT/DRE-binding factor 1 (CBF1) from *Arabidopsis Thaliana* which was related to low temperature and water deficit response [8]. The DREBs/CBFs were further divided into two subclasses, i.e., DREB1/CBF and DREB2, induced by cold and dehydration stress respectively [9]. Research on genetic engineering of plants with DREB transcription factors is still in its infancy, despite the large number of transgenic plants harboring DREBs and the initial success in achieving abiotic stress tolerance [10]. Many issues still need to be resolved to fully exploit the potential of DREB-transgenic plants under natural stress environments.

In this review, we summarize the function and regulation of DREB subfamily members in the abiotic stress responses of plants and challenges confronting deployment of DREB-transgenic plants.


## The Role of DREB in Plant Abiotic Stress Response


To date, two major schemes have been applied to define the ERF family nomenclature. The ERF and DREB were first divided into six groups in Arabidopsis [9] based on AP2/ERF domain amino acid sequences. The *Arabidopsis* and rice ERF families were then divided into 12 and 15 respective groups [11]. Similarly, 10 groups were identified in the grape and cucumber ERF family [12, 13]. In Chinese cabbage (*Brassica rapa ssp. pekinensis*), the AP2/ERF superfamily was divided into 15 groups [14].

DREB genes form a large multigene family that can be classified into six small groups named A-1 to A-6 [9]. DREB1/CBFs belong to subgroup A-1, and DREB2s belong to subgroup A-2. Later on, DREB factors belonging to other groups were identified such as ZmABI4 (A3), TINY2 (A-4), PpDBP1 (A-5), GhDBP1 (A-5), GmDREB3 (A-5), and ZmDBP1 (A-6) [15-19] [11] reported a molecular phylogenetic and motif analysis of AP2/ERF members in *Arabidopsis* and rice. They reported that A-1 and A-4 subgroups and the A-2 and A-3 subgroups described by Sakuma et al. [9] share common conserved motifs, respectively, suggesting their common origins.

Until now, most reports on DREB/CBFs transcription factors focused on DREBA1 and A2 groups, while investigation into other groups was limited.

A number of studies reported that DREB1/CBFs, DREB1A/CBF3, DREB1B/CBF1 and DREB1C/CBF2 are rapidly induced in response to cold stress [8, 20-22]. All these DREB1/CBFs contain the AP2/ERF DNA binding domain, which can recognize the CRT/DRE.

Moreover, transgenic *Arabidopsis* plants expressing DREB1B/CBF1 or DREB1A/CBF3 under the control of the CaMV 35S promoter showed a high tolerance to freezing, drought and high salinity stresses [20, 23]. In contrast, the suppression of DREB1A/CBF3 or DREB1B/CBF1 caused a reduction in freezing tolerance [24].

DREB2 genes were isolated from many economically important crops such as barley, rice, maize, wheat, sorghum, soybean, chickpea, tomato and pearl millert [25-34].

DREB2 genes, belonging to A-2, group are regulated by salt and drought, but not by cold [20, 35, 36]. However, recent reports have shown some conflicts with respect to these trends. Most of the DREB2 genes have been reported to be responsive to water stress or heat shock, while DREB2 genes from grass species were also responsive to cold stress [26, 31, 37].

DREB factors can also play a role in oxidative stress responses, as was the case for *LeDREB2* transcription factor [38]. Moreover, overexpression of AtCBF2 in *Arabidopsis* enhanced oxidative stress tolerance [39].

Recently, [40] revealed that the mechanism of NaHS-induced tolerance to salt and non-ionic osmotic stress could be via the mitigation of DREB down regulation, in accordance with other findings demonstrating that DREB2A and DREB2B were induced in Arabidopsis plants fumigated with NaHS [41].

**Table 1. DREB genes isolated from different plants and their transcript response to various abiotic stresses**

| Species | DREB type | Expression in stress | References |
|---|---|---|---|
| *Arabidopsis Thaliana* | CBF1, CBF2, CBF3 | Cold | [21] |
| *Arabidopsis Thaliana* | DREB1A, DREB2A | Cold | [20] |
| Tobacco (*Nicotina tabacum*) | Tsi1 | Salt | [42] |
| *Arabidopsis Thaliana* | CBF4 | Drought, ABA | [43] |
| Rice (*Orysa sativa)* | OsDREB1A<br>OsDREB1B<br>OsDREB1C<br>OsDREB1D<br>OsDREB2A | Cold, Salt<br>Cold<br>-<br>-<br>Drought, Salt | [35] |
| Maize (*Zea mays*) | *ZmDREB1A*<br>*ZmDREB2A* | Cold<br>Drought, heat | [44]<br>[28] |
| Wheat (*Triticum aestivum)* | TaCBF2-1<br>TaCBF2-2 | Cold, Drought<br>Cold, Drought | [45] |
| Cotton *Gossypium hirsutum* | GhDBP3 | Cold, Drought, Salt, ABA | [17] |
| Tomato *Solanum lycopersicum* | *SlDREB* | Drought | [46] |
| *Caragana korshinskii* | CkDBF | Cold, Drought, Salt, ABA | [47] |
| Soybean *Glycine max* | GmDREB2A;2<br>GmDREB 3 | Drought, Salt, Cold<br>- | [34]<br>[18] |
| Ricin *Ricinus communis L* | *RcDREB1* | Drought and salt | [48] |
| Potato *Solanum tuberosum* | *StDREB1*<br>*StDREB2* | Drought , salt, Cold and ABA | [49]<br>[50] |

## Induction of Abiotic Stress-Responsive Genes in DREB Transgenic Plants

Overexpression of specific DREB genes in transgenic plants improved plants tolerance to drought, salt and freezing stresses (Table 2).

**Table 2. DREB Transcription factors confer abiotic stress tolerance in transgenic plants**

| Transgene | Plant | Tolerance | Promoter | References |
|---|---|---|---|---|
| *AtDREB1A* *AtDREB2A* | *Arabidopsis Thaliana* | Freezing and dehydration - | *CaMV35S* | [20] |
| *AtDREB1A* | *Arabidopsis Thaliana* | Drought, salt, Freezing | *CaMV35S* | [6] [51] |
| *AtDREB1A* | *Nicotiana tabacum* | freezing, Dehydration | *CaMV35S/ RD29A* | [52] |
| *AtDREB1A* | *Solanum tuberosum* | salt | *Rd29A* | [53] |
| *AtDREB1A* | *Solanum tuberosum* | salt | *RD29A* | [54] |
| *AtDREB1A* | *Solanum tuberosum* | freezing | *RD29A* | [55] |
| *AtDREB1A* | *Chrysanthemum* | Heat | *CaMV35S* | [56] |
| *AtCBF4* | *Arabidopsis* | Drought | *CaMV35S* | [43] |
| *AtDREB1A, ABF3* | *Oryza sativa* | Drought, salt | *UBi1* | [57] |
| *AtCBF1* | *Arabidopsis* | freezing | *CaMV35S* | [23] |
| *AtCBF1* | *Solanum lycopersicum* | water deficit | *CaMV35S* | [58] |
| OsDREB1A | *Arabidopsis* | Drought, salt, freezing | *CaMV35S* | [35] |
| *StEREBP1* | *Solanum tuberosum* | cold, salt | *CaMV35S* | [59] |
| CkDBF | Tobacco | salt and osmotic stress | *CaMV35S* | [47] |
| ZmDBP2 | *Arabidopsis* | salt | *CaMV35S* | [60] |
| *GmDREB1* | *Medicago sativa* | salt | *RD29A* | [61] |
| LcDREB3a | *Arabidopsis* | Drought and salt | *CaMV35S* | [62] |
| *StDREB1* | *Solanum tuberosum* | salt | *CaMV35S* | [49] |
| *StDREB2* | *Solanum tuberosum* | salt | *CaMV35S* | [50] |

DREB transcription factors show differential transcript regulation in response to different stresses [63].

The role of DREB transcription factors in abiotic stress response was demonstrated via their overexpression in transgenic plants, the production of knockout mutants and biochemical analyses.

The constitutive over expression of AtDREB1A in transgenic *Arabidopsis* induced strong expression of *RD29A* downstream stress-responsive gene under unstressed conditions, but also as enhanced freezing and dehydration tolerance [20]. [6] Tested whether the over-expression of the AtDREB1A gene enhanced the tolerance to dehydration, they allowed the wild-type and transgenic *Arabidopsis* plants to grow in pots without watering for 2 weeks. Nearly all the wild-type plants died within this period, whereas 69.2% or even more of the 35S:DREB1A transgenic plants survived. These plants continued to grow when watering started again. Similar results were obtained for AtDREB1A-transgenic tobacco [52]; rice [57]; tall fescue [64]; and HsDREB1A-transgenic bahiagrass [65]. Moreover, the expression of AtDREB1A in transgenic *chrysanthemum* improved drought and salt tolerance of these plants that showed high proline content and SOD activity [66].

Ectopic expression of OsDREB1A in *Arabidopsis Thaliana* imparted high salt and freezing stress tolerance and also activated the over expression of several stress inducible genes such as COR15, RD29A, RD17, AtGo1S3,FL05-21, F13, FL05-20-N18 [35] .

Moreover, overexpression of rice OsDREB1F gene increased salt, drought, and low temperature tolerance in both *Arabidopsis* and rice [67].

Overexpression of AtCBF1, AtCBF3 and AtCBF4 in transgenic *Arabidopsis* up-regulated the expression of many COR genes such as COR15a, COR6.6 and COR78a [23, 43, 68].

The overexpression of AtCBF3 in *Arabidopsis* also increased freezing tolerance and, more interestingly, it led to multiple biochemical changes associated with cold acclimatization: by production of elevated levels of proline and total soluble sugars, including sucrose, raffinose, glucose, and fructose [68]. Transgenics *Arabidopsis Thaliana* plants overexpressing CBF3 exhibited high level of δ1-pyrroline-5-carboxylate synthetase (P5CS) transcripts level, suggesting that the increase in proline levels resulted from increased expression of the key proline biosynthetic enzyme P5CS [68].

Transgenic potato plants overexpressing StDREB1 or StDREB2 genes showed also higher proline levels than wild type plants, suggesting that StDREB1 and StDREB2 transcription factors activated P5CS gene transcription under salt stress conditions [50, 49]. [50] showed also that purified StDREB2 factor binds to the DRE/CRT of StP5CS gene.

Many other reports proved the role of proline in abiotic stress response through the DREB pathway. For example, tomato overexpressing AtDREB1B/CBF1 [58] and rice overexpressing OsDREB1 or AtDREB1 [69] have been shown to accumulate higher levels of proline than wild-type plants under both normal and water-deficit conditions.

Recently, the overexpression of PgDREB2A transcription factor in tobacco plants enhanced their tolerance to abiotic stresses by activating downstream stress-responsive genes such as Early Responsive to Dehydration (ERD10B), Heat shock protein (HSP70-3, Hsp18p, HSF2), Tyramine N-Hydroxycinnamoyl Transferase (THT1), and pathogen-regulated (NtERF5) factors [63].

Similar induction of stress-related genes was reported in some other transgenic plants overexpressing DREB transcription factors [56, 70-72]. [73] Reported 41 target genes of CBFs/DREBs using Affymetrix Gene Chip arrays, and showed that most of these target genes contained the DRE or DRE-related core motifs in their promoter regions. Some products of these genes belong to carbohydrate metabolism-related proteins (SPS, SuSy), prolin biosynthetic genes (P5CS), COR protein synthesis [64,68], transcription factors, phospholipase C, RNA binding proteins, sugar transport proteins, desaturases, late embryo abundant (LEA) proteins, KIN (cold inducible) proteins, molecular chaperones, osmoprotectant bio-synthesis protein, protease inhibitors, and so on [69, 74].

DREB can interact with other transcriptions factors such as WRKY, that are shown to be linked plant biotic stress response [75-83]. WRKY genes have been found also to be responsive to abiotic stresses; however, their roles in abiotic stress tolerance are largely unknown especially in crops. Recently, [84] showed that overexpression of TaWRKY19 wheat WRKY transcription factor could regulate drought and salt stress tolerance in *Arabidopsis* by activating downstream target genes related to drought resistance, such DREB2A, RD29A, RD29B and COR6.6. They revealed that and bound to DREB2A and COR6.6 promoters.


## Gene Regulation of DREB Transcription Factors

Some reports showed that the constitutive expression of DREB1/CBF genes can cause growth inhibition in *Arabidopsis* [20, 21], tobacco [52] and rice [69]. The mechanism of the growth inhibition caused by DREB1/CBF expression involves down regulation of a gibberellin (GA) signalling pathway

[85]. Indeed this report showed that the expression of DREB1B/CBF1 reduced the level of active GA by stimulating the expression of genes encoding a GA-inactivating enzyme.

DREB1/CBF transcription factors were frequently related to an ABA-independent pathway regulation [63].

Recently, DREB1/CBF regulation was also related to the circadian clock (Figure 1). Moreover the perturbation of $Ca^{2+}$ homeostasis by mutations in a vacuolar $Ca^{2+}/H^+$ antiporter gene, CALCIUM EXCHANGER 1 (CAX1), positively affected cold inducibility of DREB1/CBF and its downstream genes and then enhanced acquired freezing tolerance [86].

The *DREB1/CBF* genes were also regulated by basic helix-loop-helix bHLH-type of transcription factor, *ICE1* [87], and by $Ca2^+$ related processes, because mutations in *CAX1* (encoding a $Ca^{2+}/H^+$ transporter) and ($Ca^{2+}$ -sensor protein) *CBL1* affected expression pattern of *DREB1/CBF* genes [86, 88].

The mechanism of activation of DREB2-type genes is not well studied, indeed the overexpression of AtDREB2A and OsDREB2A protein in *Arabidopsis* was not sufficient for the induction of target stress inducible genes expression [20, 35], and it was assumed that some post-translational modifications, probably phosphorylation and/or dephosphorylation events may be necessary to play a role in activating the expression of stress responsive genes.

The DREB2A and DREB2B proteins seem to be major that are involved in dehydration inducible gene expression through DRE/CRT in an ABA-independent pathway [9]. In contrast to DREB1s/CBFs, ectopic overexpression of DREB2A didn't cause any significant changes in plant growth, gene expression or stress tolerance [20].

Phosphorylation has been suggested as a post-translational regulatory mechanism that is necessary for the activation of DREB2 transcription factors under drought stress conditions [89]. [27] have shown for the first time that PgDREB2A is a phosphotrotein and its DNA binding is negatively influenced by phosphorylation and the maximum induction was seen by low temperature. Moreover, AtDREB2A and OsDREB2A transcription factors contain a conserved serine/threonine-rich region adjacent to the AP2/ERF domain, with putative sites of phosphorylation [20, 36]. The overexpression of both genes in *Arabidopsis* induced a weak expression of the target genes. The authors suggest that not only transcriptional regulation but also post-translational modification may be required for the activation of DREB2A proteins under drought stress conditions. In this context, [20] reported the requirement of post-translational modifications, probably phosphorylation/dephosphorylation for activation of DREB2A. These post-translational modifications may require cofactors, which are dependent on ABA-regulated molecules such as ABI1, ABI2, $Ca^{2+}$ dependent protein kinases, or other ABA responsive regulatory factors [90, 91].

In *Arabidopsis*, DREB2A and DREB2B are induced by dehydration, they seem to be major DREB2 factors involved in dehydration inducible gene expression through DRE/CRT in an ABA-independent pathway [9, 36].

The dehydration inducible genes contain potential Abscisic acid Responsive Elements-ABREs (PyACGTGGC) in their promoter region. The ABRE functions as a cis acting DNA element involved in ABA regulated gene expression.

A coupling element (CE3) is needed to specify the function of ABRE for the expression of ABA induced genes [92].

Furthermore, it has been reported that DREB2A promoter contains the ABRE element and the coupling element 3 (CE3). These sequences are necessary for the promoter activity under stress conditions such as osmotic, dehydration and high salinity [93].

A number of studies indicated that the ABA-dependent and ABA-independent pathways have extensive interactions in controlling gene expression under abiotic stress conditions [94, 95]. Recently, [96] reviewed that there are a cross-talk between abscisic acid-dependent and abscisic acid-independent pathways during abiotic stress response.

Some DREB genes, such as barley HvDRF1 and wheat TaDREB2 and TaAIDFa exhibited responses to exogenous ABA [25, 26, 97]. ABA was shown to be involved in the regulation of DREB activity and increased promoter activity [15, 97].

DREB1A/CBF3, DREB2A, and DREB2C proteins have been reported to interact physically with AREB/ABF proteins [98], which supports the view that DREB/CBFs and AREB/ABFs may interact to control ABA-regulated gene expression.

The *RD29A* promoter has both DREs (three) and ABREs (one) sequences. The ABA-independent DREB genes are expressed rapidly (within 20 min) in response to dehydration, cold and salinity stress [99, 100], but not by ABA as ABA biosynthesis gets induced after 2 h of dehydration and high salinity stresses [101]. In contrast, the ABRE functions in induction of *RD29* after the accumulation of ABA under dehydration and high salt conditions. Since both DRE and ABRE mediate the expression of *RD29* gene, an interaction between these two signaling pathways is needed to maintain its level during different stress response.

Furthermore, DREB2 was also shown to be a component of a transcriptional cascade involved in heat-shock responses [102].

A number of heat shock related genes (268 and 778 at 0.5 h and 5 h heat shock treatment) were upregulated by *AtDREB2A* [103]. [104] reported that HEAT SHOCK TRANSCRIPTION FACTOR A3 (HsfA3), one of the 21 members of *Arabidopsis* Hsf family, was transcriptionally activated by DREB2A during heat shock. DREB2A then activates the promoter of HsfA3, and the expressed HsfA3 protein further induces downstream genes (Figure 1).

DREB factors interact with other transcription factors and DRE/CRT interacts with other cis-acting elements (Figure 1).

DRE/CRT and heat shock elements (HSEs) cooperatively function in the promoter of a small heat shock protein during development sunflower embryos [105]. Similarly, in the RD29A promoter, DRE/CRT had a positive relationship with ABRE and functions as a coupling element of ABRE [106].

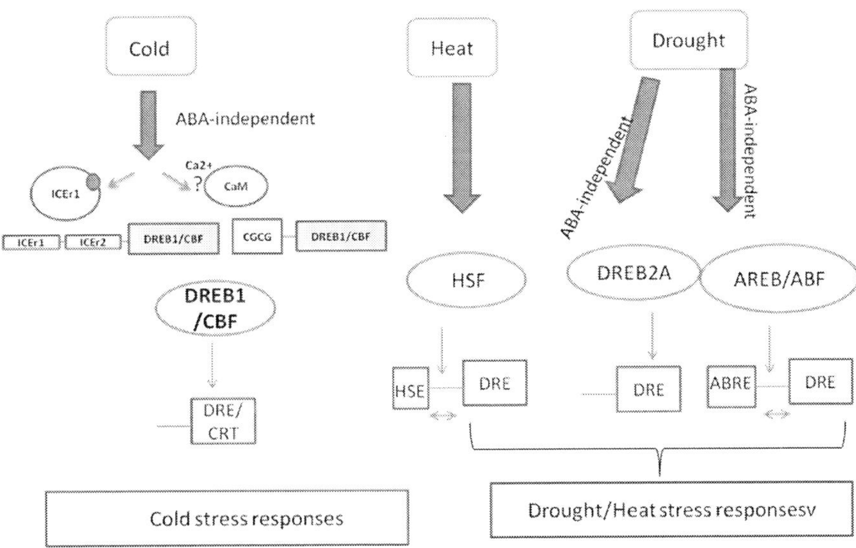


Figure 1. Model showing DREB transcription factors involved in regulation of expression of stress responsive genes under various abiotic stresses through different pathways.

The activity of DREB1s/CBFs is mainly regulated at the transcriptional level. The transcription of DREB1s/CBFs is under the control of low-temperature signals and circadian/light signals. CaM: calmodulin, CGCG: CGCG-Box.

DREB2A activates expression of target genes in a stress-specific manner via DRE/CRT sequences in the promoters. Heat- and drought-inducible genes' promoters often contain heat shock elements (HSEs) and ABA-responsive elements (ABREs), respectively. Interaction between DRE/CRT and these cis-elements has been suggested. Such interactions may involve interactions between transcription factors (Mizoi et al. 2012).

# CONCLUSION

DREB1/CBF and DREB2 subgroups play central roles in the acquisition of stress tolerance by regulating the expression of gene sets via DRE/CRT sequences in promoters of stress-inducible genes.

The activity of DREB1s/CBFs is mainly regulated at the transcriptional level. In *Arabidopsis*, the expression of three cold-inducible DREB1s/CBFs is under the control of low temperature and circadian signals.

In contrast, the activity of DREB2s is regulated by multiple steps. In addition to the transcriptional activation, alternative splicing is important in grass species, whereas post-translational stabilisation and activation are required in Arabidopsis.

Transformation of plants with DREBs is one of the preferred strategies to develop multiple abiotic stress tolerance. From the above examples, it is evident that in response to abiotic stresses, the DREB transcription factors up-regulate the expression of many stress-related genes, the products of which work in various ways to confer protection.

In response to drought and salt stress, most of the DREB transgenic plants accumulated higher proline and soluble sugar content compared to that of wild controls. Proline is an important osmoprotectant that works in osmotic adjustment, protection of macromolecules and scavenging reactive oxygen species (ROS). Soluble sugar accumulation assists in osmotic adjustment. In addition to other protective mechanisms, accumulation of proline and soluble sugars seems to be the main defense strategy against dehydration and salt stress in DREB-transgenic plants.

In order to evaluate the increased weediness and invasiveness potential in abiotic stress tolerant transgenic plants engineered with DREBs, emphasis should be placed on the problem formulation during environmental risk assessment. Future research should focus on additional approaches including interactions between DRE/CRT and other cis-acting elements, as well as interactions between AP2/ERF transcription factors and other proteins.

These efforts should be integrated to make DREBs more effective, target specific and stress specific.

## ACKNOWLEDGMENTS

This work was financed by the Tunisian Ministry of High Education and Scientific Research. Authors are grateful to Dr. Anne- Lise Haenni from Institute Jacques Monod (France) for reading and improving the manuscript and to Mofida Bouaziz-Kanoun from the "Institut Supérieur d'Administration des Affaires de Sfax" (Tunisia) for her kind help with the English language.

## REFERENCES

[1]    Wang, W., Vinocur, B., Shoseyov, O., & Altman, A. (2001). Biotechnology of plant osmotic stress tolerance: physiological and molecular considerations. *Acta. Hortic.*, 560, 285-292.

[2]    Yamaguchi-Shinozaki, K., & Shinozaki, K. (2006). Transcriptional regulatory networks in cellular responses and tolerance to dehydration and cold stresses. *Annu. Rev. Plant. Biol.*, 57, 781-803.

[3]    Jofuku, K.D., den Boer, B.G., Van Montagu, M. & Okamuro, J.K. (1994). Control of Arabidopsis flower and seed development by the homeotic gene APETALA2. *Plant Cell,* 6, 1211-1225.

[4]    Ohme-Takagi, M. & Shinshi, H. (1995). Ethylene-inducible DNA binding proteins that interact with an ethylene responsive element. *Plant Cell,* 7, 173-182.

[5]    Gu, Y. Q., Yang, C., Thara, V., Zhou, J.-M., & Martin, G. B. (2000). Pti4 is induced by ethylene and salicylic acid, and its product is phosphorylated by the Pto kinase. *Plant Cell.* 12,771-785.

[6]    Kasuga, M., Liu, Q., Miura, S., Yamaguchi-Shinozaki, K., & Shinozaki, K. (1999).Improving plant drought, salt, and freezing tolerance by gene transfer of a single stress inducible transcription factor. *Nat. Biotechnol,* 17, 287-291.

[7]    Liu, Y., Zhao, T., Liu, J., Liu, W., Liu, Q., Yan, Y., & Zhou, H., (2006). The conserved Ala37 in the ERF/AP2 domain is essential for binding with the DRE element and the GCC box. *FEBS Letters,* 580, 1303-1308.

[8]     Stockinger, E J., Gilmour, SJ., & Thomashow, M.F., (1997) *Arabidopsis thaliana* CBF1 encodes an AP2 domain-containing transcriptional activator that binds to the C-repeat/DRE, a cis-acting DNA regulatory element that stimulates transcription in response to low temperature and water deficit. *Proc. Natl. Acad. Sci. USA* ,94, 1035-1040.

[9]     Sakuma, Y., Liu, Q., Dubouzet, J.G., Abe, H., Shinozaki, K., & Yamaguchi-Shinozaki, K., (2002) DNA-binding specificity of the ERF/AP2 domain of Arabidopsis DREBs, transcription factors involved in dehydration and coldinducible gene expression. *Biochem. Biophys. Res. Commun*, 290, 998-1009.

[10]   Sayyar Khan, M., (2011). The role of DREB transcription factors in abiotic stress tolerance of plants. *Agriculture and environmental biotechnology.* DOI: 10.5504/bbeq.2011.0072.

[11]   Nakano, T., Suzuki, K., Fujimura, T., & Shinshi, H., (2006) Genome-wide analysis of the ERF gene family in Arabidopsis and rice. *Plant Physiol.,* 140, 411-432.

[12]   Licausi, F., Giorgi, F.M., Zenoni, S., Osti, F., Pezzotti, M., & Perata, P., (2010). Genomic and transcriptomic analysis of the AP2/ERF superfamily in *Vitis vinifera. BMC Genomics*, 11,719.

[13]   Hu, L., & Liu, S., (2011). Genome-wide identification and phylogenetic analysis of the ERF gene family in cucumbers. *Genet. Mol. Biol.,* 34(4),624-633.

[14]   Song, X., Li, Y., & Hou, X., (2013). Genome-wide analysis of the AP2/ERF transcription factor superfamily in Chinese cabbage (*Brassica rapa ssp. pekinensis*) *BMC Genomics*, 14:573.

[15]   Kizis, D., & Pages, M., (2002). Maize DRE-binding proteins DBF1 and DBF2 are involved in rab17 regulation through the drought responsive element in an ABA-dependent pathway. *Plant J.* 30, 679-689.

[16]   Wei, G., Pan, Y., Lei , J., & Zhu, Y.X. (2005). Molecular cloning, phylogenetic analysis, expressional prowling and in vitro studies of TINY2 from Arabidopsis thaliana. *J. Biochem. Mol. Biol.* 38, 440-446.

[17]   Huang, B., & Liu, J.Y., (2006). A cotton dehydration responsive element binding protein functions as a transcriptional repressor of DRE-mediated gene expression, *Biochem. Biophys. Res. Commun.* 343, 1023-1031.

[18]   Nasreen, S., Amudha, J., & Pandey, S.S. (2013). Isolation and characterization of Soybean DREB 3 transcriptional activator *Journal of Applied Biology & Biotechnology* 1 (02), 009-012.

[19]   Liu, N., Zhong, N.Q., Wang, G. L., Li, L.J. , Liu, X. L., He, Y.K., & Xia, G.X. (2007). Cloning and functional characterization of PpDBF1

gene encoding a DRE-binding transcription factor from Physcomitrella patens, *Planta* 226, 827-838.

[20] Liu, Q., Kasuga, M., Sakuma, Y., Abe, H., Miura, S., Yamaguchi-Shinozaki, K., & Shinozaki, K. (1998) Two transcription factors, DREB1 and DREB2, with an EREBP/AP2 DNA binding domain separate two cellular signal transduction pathways in drought and low temperature responsive gene expression, respectively, in Arabidopsis. *The Plant Cell* , 10, 1391-1406.

[21] Gilmour, S.J., Zarka, D.G., Stockinger, E.J., Salazar, M.P., Houghton, J.M., & Thomashow, M.F., (1998). Low temperature regulation of the Arabidopsis CBF family of AP2 transcriptional activators as an early step in cold induced COR gene expression. *Plant Journal*, 16, 433-442.

[22] Shinwari, Z.K., Nakashima, K., Miura, S., Kasuga, M., Seki, M., Yamaguchi-Shinozaki, K. & Shinozaki, K., (1998). An Arabidopsis gene family encoding DRE/CRT binding proteins involved in low-temperature-responsive gene expression. *Biochem. Biophys. Res. Commun.* 250, 161-170.

[23] Jaglo-Ottosen, K.R., Gilmour, S.J., Zarka, D.G., Chabenberger, O.S & Thomashow, M.F., (1998). *Arabidopsis* CBF1 overexpression induces COR genes and enhances freezing tolerance. *Science*, 280, 104-106.

[24] Novillo, F., Medina, J., & Salinas, J., (2007) Arabidopsis CBF1 and CBF3 have a different function than CBF2 in cold acclimation and define different gene classes in the CBF regulon. *Proc. Natl. Acad. Sci. U. S. A.* 104 , 21002–21007.

[25] Xue, G.P., & Loveridge, C.W., HvDRF1 is involved in abscisic acid-mediated gene regulation in barley and produces two forms of AP2 transcriptional activators, interacting preferably with a CT-rich element, *Plant J.* 37 (2004) 326-339.

[26] Egawa, C., Kobayashi, F., Ishibashi, M., Nakamura, T., Nakamura, C., & Takumi, S. (2006) Differential regulation of transcript accumulation and alternative splicing of a DREB2 homolog under abiotic stress conditions in common wheat, *Genes. Genet. Syst.* 81, 77-91.

[27] Agarwal, P., Agarwal, P., Nair, S., Sopory, S., & Reddy, M., (2007) Stress-inducible DREB2A transcription factor from Pennisetum glaucum is a phosphoprotein and its phosphorylation negatively regulates its DNA-binding activity, *Mol. Genet. Genomics* 277 189-198.

[28] Qin, F., Kakimoto, M., Sakuma, Y., Maruyama, K. Osakabe, Y., Tran, L.S.P., Shinozaki, K. & Yamaguchi-Shinozaki, K. (2007). Regulation

and functional analysis of ZmDREB2A in response to drought and heat stresses in *Zea mays* L, *Plant J.* 50, 54-69.

[29] Nayak , S. N., Balaji, J., Upadhyaya, H. D., Hash, C. T., Kishor, P.B. K.,Chattopadhyay, D ., Rodriquez, L.M., Blair, M. W., Baum, M., McNally, K ., This, D., Hoisington, D. A., & Varshney, R. K., (2009). Isolation and sequence analysis of DREB2A homologues in three cereal and two legume species. *Plant Science*, 177, (5), 460-467

[30] Bihani, P., Char, B., & Bhargava, S., (2011). Transgenic expression of sorghum DREB2 in rice improves tolerance and yield under water limitation. *The Journal of Agricultural Science* 149, 95-101.

[31] Matsukura, S. Mizoi, J. Yoshida, T. Todaka, D., Ito, Y., Maruyama, K., Shinozaki, K., & Yamaguchi-Shinozaki, K., (2010). Comprehensive analysis of rice DREB2-type genes that encode transcription factors involved in the expression of abiotic stress-responsive genes, *Mol. Genet. Genomics,* 283, 185-196.

[32] Morran, S., Eini, O., Pyvovarenko, T., Parent, B., Singh, R., Ismagul, A., Eliby, S., Shirley, N., Langridge, P., & Lopato, S., (2011). Improvement of stress tolerance of wheat and barley by modulation of expression of DREB/CBF factors. *Plant Biotechnol.* J ,9 (2),230-49.

**[33]** Mallikarjuna, G., Mallikarjuna, K., Reddy, M. K., & Kaul, T., (2011). Expression of OsDREB2A transcription factor confers enhanced dehydration and salt stress tolerance in rice (*Oryza sativa* L.) *Biotechnology Letters* 33, (8) 1689-1697.

[34] Mizoi, J., Ohori, T., Moriwaki, T., Kidokoro, S., Todaka, D., Maruyama, K., Kusakabe, K., Osakabe, Y., Shinozaki, K., & Yamaguchi-Shinozaki, K (2013). GmDREB2A;2, a Canonical DEHYDRATION-RESPONSIVE ELEMENT-BINDING PROTEIN2-Type Transcription Factor in Soybean, Is Posttranslationally Regulated and Mediates Dehydration-Responsive Element-Dependent Gene Expression .*Plant Physiology* , 161, 346-361

[35] Dubouzet, J.G., Sakuma, Y., Ito, Y., Kasuga, M., Dubouzet, E.G., Miura, S., Seki, M. Shinozaki, K., & Yamaguchi-Shinozaki K., (2003). OsDREB genes in rice, Oryza sativa L., encode transcription activators that function in drought-, high-salt- and cold-responsive gene expression, *Plant J.* 33, 751-763.

[36] Nakashima, K. Shinwari, Z.K. Sakuma, Y. Seki, M. Miura, S. Shinozaki, K., & Yamaguchi-Shinozaki, K. (2000). Organization and expression of two Arabidopsis DREB2 genes encoding DRE-binding proteins

involved in dehydration- and highsalinity- responsive gene expression. *Plant Mol. Biol*, 42, 657-665.

[37] Shen, Y.G, Zhang, W.K., He, S.J., Zhang, J.S., Liu, Q., & Chen, S.Y., (2003). An EREBP/AP2-type protein in Triticum aestivum was a DRE-binding transcription factor induced by cold, dehydration and ABA stress, *Theor. Appl. Genet*, 106, 923-930.

[38] Guo, J., & Wang, M.H., (2011). Expression profiling of the DREB2 type gene from tomato (*Solanum lycopersicum L.*) under various abiotic stresses. *Horticulture, Environment, and Biotechnology* 52 (1), 105-111

[39] Sharabi-Schwager, M., Porat, R., & Samach, A., (2011).Overexpression of the CBF2 transcriptional Activator Enhances Oxidative Stress Tolerance in Arabidopsis Plants. *International Journal of Biology* 3, (2)

[40] Christou, A., Manganaris, G.A., Papadopoulos, I. & Fotopoulos, V., (2013). Hydrogen sulfide induces systemic tolerance to salinity and non-ionic osmotic stress in strawberry plants through modification of reactive species biosynthesis and transcriptional regulation of multiple defence pathways. *Journal of Experimental Botany*, 64 (7) 1953-1966

[41] Jin, Z., Shen, J., Qiao, Z., Yang, G., Wang, R., & Pei, Y., (2011). Hydrogen sulfide improves drought resistance in *Arabidopsis thaliana*. *Biochemical and Biophysical Research Communications* 414, 481-486.

[42] Park, J.M., Park, C.J., Lee, S.B., Ham, B.K., Shin, R., & Paek, K.H. (2001) Overexpression of the tobacco Tsi1 gene encoding an EREBP/AP2-type transcription factor enhances resistance against pathogen attack and osmotic stress in tobacco." Plant Cell 13,1035-1046.

[43] Haake, V., Cook, D., Riechmann, J.L., Pineda, O., Thomashow, M.F., & Zhang, J.Z. (2002).Transcription factor CBF4 is a regulator of drought adaptation in Arabidopsis. *Plant Physiol.* 130, 639-648.

[44] Qin, F., Sakuma, Y., Li, J., Liu, Q., Li, Y.Q., Shinozaki, K., &Yamaguchi-Shinozaki K., (2004). Cloning and functional analysis of a novel DREB1/CBF transcription factor involved in cold-responsive gene expression in *Zea mays L. Plant Cell. Physiol.*,45,1042-1052.

[45] Kume, S., Kobayashi, F., Ishibashi, M., Ohno, R., Nakamura, C., & Takumi, S., (2005). Differential and coordinated expression of Cbf and Cor/ Lea genes during long-term cold acclimation in two wheat cultivars showing distinct levels of freezing tolerance. *Genesand & Genetic Systems* 80, 185-197.

[46] Li, J., Sima, W., Ouyang, B., Wang, T., Ziaf, K., Luo, Z., Liu, L., Li, H., Chen, M., Huang, Y., Feng, Y., Hao, Y., & Ye, Z., (2012). Tomato SlDREB gene restricts leaf expansion and internode elongation by

downregulating key genes for gibberellins biosynthesis. *Journal of Experimental Botany*, 63, 18, 6407-6420.

[47] Wang, X., Dong, J., Liu, Y., & Gao, H. (2010). A novel dehydration-responsive element-binding protein from Caragana korshinskii is involved in the response to multiple abiotic stresses and enhances stress tolerancein transgenic tobacco. *Plant Molecular Biology Reporter*, 28, 664-675.

[48] Cipriano, T.M., Morais, A.T., Aragão, F. J. L. (2013). Characterization of a pollen-specific and desiccation-associated AP2/ERF type transcription factor gene from castor bean (*Ricinus communis* L.). *International Journal of Plant Biology*, 4, 1.

[49] Bouaziz, D., Pirrello, J., Charfeddine, M., Hammami, A., Jbir, R., Dhieb, A., Bouzayen, M., & Gargouri-Bouzid, R., (2013) Overexpression of StDREB1 Transcription Factor Increases Tolerance to Salt in Transgenic Potato Plants. *Mol. Bio.* 54 (3), 803-817.

[50] Bouaziz, D., Pirrello, J., Ben Amor, H., Hammami, A., Charfeddine, M., Dhieb, A. Bouzayen, M., & Gargouri-Bouzid, R., (2012) Ectopic expression of dehydration responsive element binding proteins (StDREB2) confers higher tolerance to salt stress in potato. *Plant. Physiol. Biochem.* 60, 98-108.

[51] Yamaguchi-Shinozaki, K., Shinozaki, K., (2001). Improving plant drought, salt and freezing tolerance by gene transfer of a single stress-inducible transcription factor. *Novartis Found Symp.* 236,176-186.

[52] Kasuga, M., Miura, S., Shinozaki, K. & Yamaguchi-Shinozaki, K. (2004). A combination of the Arabidopsis DREB1A gene and stress-inducible rd29A promoter improved drought- and low-temperature stress tolerance in tobacco by gene transfer, *Plant Cell Physiol.* 45, 346-350.

[53] Celebi-Toprak, F., Behnam, B., Serrano, G., Kasuga, M., Yamaguchi-Shinozaki, K., Naka, H., & et al. (2005). Tolerance to salt stress of the transgenic tetrasomic tetraploid potato, Solanum tuberosum cv. Desiree appears to be induced by the DREB1A gene and rd29A promoter of Arabidopsis thaliana. *Breeding Science*, 55, 311-319.

[54] Behnam, B., Kikuchi, K., Celebi-Toprak, F., Yamanaka, S., Kasuga, M., Yamaguchi-Shinozaki, K., et al. (2006). The Arabidopsis DREB1A gene driven by the stress-inducible rd29A promoter increases salt-stress tolerance in tetrasomic tetraploid potato (*Solanum tuberosum*) in proportion to its copy number. *Plant Biotechnology*, 23, 169-177.

[55] Behnam,B., Kikuchi,A., Celebi-Toprak, F., Kasuga,M., Yamaguchi-Shinozaki, K., & Watanabe, K. N. (2007). Arabidopsis rd29A: DREB1A

enhances freezing tolerance in transgenic potato. *Plant Cell Reports*, 26, 1275-1282.

[56] Hong, B., Ma, C., Yang, Y., Wang, T., Yamaguchi- Shinozaki, K. & Gao J. (2009). Over- expression of AtDREB1A in chrysanthemum enhances tolerance to heat stress. *Plant Molecular Biology* 70, 231-240.

[57] Oh, S.J., Song, S.I., Kim, Y.S., Jang, H.J., Kim, S.Y., Kim, M.J., Kim, Y.K., Nahm, B.H., & Kim, J.K. (2005). Arabidopsis CBF3/DREB1A and ABF3 in Transgenic Rice Increased Tolerance to Abiotic Stress without Stunting Growth Plant. *Physiol.* 138, 341-351.

[58] Hsieh, T.H., Lee, J.T., Yang, P.T., Chiu, L.H., Charng, Y.Y., Wang, Y.C., & Chan, M.T., (2002). Heterology expression of the Arabidopsis C-repeat/dehydration response element binding factor 1 gene confers elevated tolerance to chilling and oxidative stresses in transgenic tomato, *Plant Physiol.* 129, 1086-1094.

[59] Lee, H.E., Shin, D., Park, S.R., Han, S.E., Jeong, M.J., Kwon, T.R., Lee, S.K., Park, S.C., Yi, B.Y., Kwon, H.B., Byun, M.O., (2007). Ethylene responsive element binding protein 1 (StEREBP1) from Solanum tuberosum increases tolerance to abiotic stress in transgenic potato plants. *Biochem. Biophys. Res. Commun.* 353,863-868.

[60] Wang, C. T., Yang, Q., & Wang, C. T., (2011). Isolation and functional characterization of ZmDBP2 encoding a dehydrationresponsive element-binding protein in zea mays. *Plant Molecular Biology Reporter*, 29, 60-66.

[61] Jin, T., Chang, Q., Li, W., Yin, D., Li, Z., Wang, D., Liu, B., Liu, L. (2010).Stress-inducible expression of GmDREB1 conferred salt tolerance in transgenic alfalfa. *Plant Cell Tissue Organ Cult* 100, 219-227.

[62] Xianjun, P., Xingyong, M., Weihong, F., Man, S., Liqin, C., Alam, I., Lee, B.H., Dongmei, Q., Shihua, S., Gongshe, L., (2011). Improved drought and salt tolerance of Arabidopsis thaliana by transgenic expression of a novel DREB gene from Leymus chinensis. *Plant Cell Rep.*30(8),1493-502.

[63] Agarwal, P., Agarwal, P.K., Joshi, A.J., Sopory, S.K. & Reddy, M.K. (2010) Overexpression of PgDREB2A transcription factor enhances abiotic stress tolerance and activates downstream stress-responsive genes. *Mol. Biol. Rep.* 37, 1125-1135.

[64] Zhao, J., Ren, W., Zhi, D., Wang, L., & Xia, G., (2007). Arabidopsis *DREB1A/CBF3* bestowed transgenic tall fescue increased tolerance to drought stress. *Plant Cell Reports.* 26, (9), 1521-1528.

[65] James, V.A., Neibaur, I., & Altpeter, F., (2008). Stress inducible expression of the DREB1A transcription factor from xeric, Hordeum spontaneum L. in turf and forage grass (Paspalum notatum Flugge) enhances abiotic stress tolerance. *Trans. Res.*, 17, 93-104.

[66] Hong, B., Tong, Z., Ma, N., Kasuga, M., Yamaguchi-Shinozaki, K., & Gao, J.P., (2006). Expression of Arabidopsis DREB1A gene in transgenic chrysanthemum enhances tolerance to low temperature. *J. Hortic. Sci. Biot.*, 81, 1002-1008.

[67] Wang, Q., Guan, Y., Wu, Y., Chen, H., Chen, F., & Chu, C., (2008). Overexpression of a rice OsDREB1F gene increases salt, drought, and low temperature tolerance in both Arabidopsis and rice. *Plant mol. Biol.* 67, 589-602.

[68] Gilmour, S.J., Sebolt, A.M., Salazar, M.P., Everard, J.D., & Thomashow, M.F., (2000). Over-expression of the Arabidopsis CBF3 transcriptional activator mimics multiple biochemical changes associated with cold acclimation. *Plant Physiol.*, 124, 1854-1865.

[69] Ito, Y., Katsura, K., Maruyama, K., Taji, T., Kobayashi, M., Seki, M., Shinozaki, K. & Yamaguchi-Shinozaki, K., (2006). Functional analysis of rice DREB1/CBF-type transcription factors involved in cold-responsive gene expression in transgenic rice, *Plant Cell Physiol.* 47, 141-153.

[70] Ashraf, M., & Akram, N.A., (2009). Improving salinity tolerance of plants through conventional breeding and genetic engineering: an analytical comparison. *Biotechnol. Adv.*, 27, 744-52.

[71] Chen, M., Wang, Q.Y., Cheng, X.G., Xu, Z.S., Li, L.C., Ye, X.G., Xia, L.Q., & Ma, Y.Z.(2007). GmDREB2, a soybean DRE-binding transcription factor, conferred drought and high-salt tolerance in transgenic plants. *Biochem. Bioph. Res. Co.*, 353, 299-305.

[72] Cong, L., Chai, T.Y., & Zhang, Y.X., (2008). Characterization of the novel gene BjDREB1B encoding a DRE-binding transcription factor from *Brassica junceaL. Biochem. Biophys. Res. Commun.* 371, 702-706.

[73] Fowler, S., & Thomashow, M.F., (2002). *Arabidopsis* transcriptome profiling indicates that multiple regulatory pathways are activated during cold acclimation in addition to the CBF cold response pathway. *Plant Cell.* 14, 1675-1690.

[74] Maruyama, K., Sakuma, Y., Kasuga, M., Ito, Y., Seki, M., Goda, H., Shimada, Y., Yoshida, S., Shinozaki, K., & Yamaguchi-Shinozaki, K., (2004). Identification of cold-inducible downstream genes of the

ArabidopsisDREB1A/CBF3 transcriptional factor using two microarray systems. *The Plant Journal* 38, 982-993.

[75] Yang, Y., & Klessig, D.F. (1996). Isolation and characterization of a tobacco mosaic virus–inducible *myb* oncogene homolog from tobacco. *Proc. Natl. Acad. Sci. USA* 93, 14972-14977.

[76] Zhou, J., Tang, X., & Martin, G.B. (1997). The Pto kinase conferring resistance to tomato bacterial speck disease interacts with proteins that bind a *cis*–element of pathogenesis–related genes. *EMBO J.* 16, 3207-3218.

[77] Rushton, P.J., & Somssich, I.E., (1998). Transcriptional control of plant genes responsive to pathogens. *Curr. Opin. Plant Biol.* 1, 311-315.

[78] Eulgem, T., Rushton, P.J., Robatzek, S., & Somssich, I.E., (2000). The WRKY superfamily of plant transcription factors. *Trends Plant Sci.* 5, 199-206.

[79] Jakoby, M., Weisshaar, B., Dröge–Laser, W., Vicente–Carbajosa, J., Tiedemann, J., Kroj, T., & Parcy, F. (2002). bZIP transcription factors in Arabidopsis. *Trends Plant Sci.* 7, 106-111.

[80] Desveaux, D., Subramaniam, R., Després, C., Mess, J.N., Lévesque, C., Fobert, P.R., Dangl, J.L., & Brisson, N. (2004). A Whirly transcription factor is required for salicylic acid–dependent disease resistance in *Arabidopsis. Dev. Cell.* 6, 229-240.

[81] Ülker, B., & Somssich, I.E., (2004). WRKY transcription factors: from DNA binding towards biological function. *Curr. Opin. Plant Biol.* 7, 491-498.

[82] Turck, F., Zhou, A., & Somssich, I.E., (2004). Stimulus–dependent, promoter–specific binding of transcription factor WRKY1 to its native promoter and the defense–related gene *PcPR1–1* in parsley. *Plant Cell* 16, 2573-2585.

[83] Desveaux, D., Maréchal, A., & Brisson, N. (2005). Whirly transcription factors: defense gene regulation and beyond. *Trends Plant Sci.* 10, 95-102.

[84] Niu, C.F, Wei, W., Zhou, Q.Y, Tian, A.G, Hao, Y.J, Zhang, W.K, Ma, B., Lin, Q., Zhang, B,Z., Zhang, J.S., & Chen, S.Y., (2012).Wheat WRKY genes TaWRKY2 and TaWRKY19 regulate abiotic stress tolerance in transgenic *Arabidopsis plants Plant, Cell & Environment* 35, (6) 1156-1170.

[85] Achard, P., Gong, F., Cheminant, S., Alioua, M., Hedden, P., & Genschik, P., The cold inducible CBF1 factor-dependent signaling pathwaymodulates the accumulation of the growth-repressing DELLA

proteins via its effect on gibberellin metabolism, *Plant Cell* 20 (2008) 2117-2129.

[86] Catalá, R., Santos, E., Alonso, J.M., Ecker, J.R., Martínez-Zapater, J.M., & Salinas, J., (2003). Mutations in the Ca2+/H+ transporter CAX1 increase CBF/DREB1 expression and the cold-acclimation response in Arabidopsis, *Plant Cell* 15, 2940-2951

[87] Chinnusamy. V., Ohta, M., Kanrar, S., Lee, B.H., Hong, X., Agarwal, M., & Zhu, J. K., (2003) ICE1 a regulator of cold-induced transcriptome and freezing tolerance in Arabidopsis, *Genes Dev.* 17, 1043-1054.

[88] Albrecht, V., Weinl, S., Blazevic, D., D'Angelo, C., Batistic, O., Kolukisaoglu, U., Bock, R., Schulz, B., Harter, K., & Kudla, J., (2003). The calcium sensor CBL1 integrates plant responses to abiotic stresses. *Plant J* 36,457-470.

[89] Busk, P.K., & Pages, M., (1998). Regulation of abscisic acid-induced transcription. *Plant Mol. Biol.* 37, 425-435.

[90] Leung, J., Merlot, S., & Giraudat, J., (1997). The Arabidopsis ABSCISIC ACID–INSENSITIVE2 (ABI2) and ABI1 genes encode homologous protein phosphatase 2C involved in abscisic acid signal transduction. *Plant Cell* 9, 759-771.

[91] Merlot, S., Costi, F., Guerrier, D., Vavasseur, A., & Giraudat, J. (2001).The ABI1 and ABI2 protein phosphatases 2C act in a negative feedback regulatory loop of the abscisic acid signaling pathway. *Plant J.* 25, 295-303.

[92] Shen, Q., & Ho, T.H.D. (1995). Functional dissection of an abscisic acid (ABA)-inducible gene reveals two independent ABAresponsive complexes each containing a G-box and a novelcis-acting element. *Plant Cell*, 7, 295-307.

[93] Kim, J.S, Mizoi, J., Yoshida, T., Fujita, Y., Nakajima, J., Ohori, T., Todaka, D., Nakashima, K., Hirayama, T., Shinozaki, K., & Yamaguchi-Shinozaki, K., (2011). An ABRE Promoter Sequence is Involved in Osmotic Stress-Responsive Expression of the DREB2A Gene, Which Encodes a Transcription Factor Regulating Drought-Inducible Genes in Arabidopsis. *Plant Cell Physiol*, 52(12), 2136 -2146.

[94] Ishitani, M., Xiong, L., Stevenson, B., & Zhu, J. K. (1997) Genetic analysis of osmotic and cold stress signal transduction in Arabidopsis: interactions and convergence of abscisic acid dependent and abscisic acid-independent pathways. *Plant Cell,* 9, 1935-1949.

[95]  Xiong, L., Ishitani, M., Lee, H., & Zhu, J.K. (1999) HOS5-a negative regulator of osmotic stress-induced gene expression in Arabidopsis thaliana. *Plant J.* (19), 569-578.

[96]  Roychoudhury, A., Paul, S., & Basu, S., (2013). Cross-talk between abscisic acid-dependent and abscisic acid-independent pathways during abiotic stress. *Plant Cell Reports*, 32, (7), 985-1006.

[97]  Xu, Z.S., Ni, Z.Y., Liu, L., Nie, L.N., Li, L.C., Chen, M., & Ma, Y.Z. (2008). Characterization of the TaAIDFa gene encoding a CRT/DRE binding factor responsive to drought, high-salt, and cold stress in wheat. *Mol. Genet. Genomics,* 280, 497-508.

[98]  Lee, S.J., Kang, J.Y., Park, H.J., Kim, M.D., Bae, M.S., Choi, H.I., & Kim, S.Y., (2010) DREB2C Interacts with ABF2, a bZIP Protein Regulating Abscisic Acid-Responsive Gene Expression, and Its Overexpression Affects Abscisic Acid Sensitivity. *Plant Physiol*, 153,716-727.

[99]  Yamaguchi-Shinozaki, K., & Shinozaki, K., (1993). Characterization of the expression of a desiccation-responsive rd29 gene of *Arabidopsis thaliana* and analysis of its promoter in transgenic plants. *Mol. Gen. Genet.* 236, 331-340.

[100]  Yamaguchi-Shinozaki K, Shinozaki K (1994) A novel cis-acting element in an Arabidopsis gene is involved in responsiveness to drought, low-temperature, or high-salt stress. *Plant Cell* 6:251–264.

[101]  Kiyosue, T., Beetham, J.K., Pinot, F ., Hammock, B.D., Yamaguchi-Shinozaki, K., & Shinozaki K. (1994). Characterization of an Arabidopsis cDNA for a soluble epoxide hydrolase gene that is inducible by auxin and water stress. *Plant J.* 6 (2), 259-69.

[102]  Mizoi, J., Shinozaki, K., & Yamaguchi-Shinozaki, K., (2012). AP2/ERF family transcription factors in plant abiotic stress responses. *Biochimica et Biophysica Acta* 1819, 86.

[103]  Sakuma, Y. Maruyama, K. Osakabe, Y. Qin, F. Seki, M. Shinozaki, K. Yamaguchi- & Shinozaki, K. (2006) Functional analysis of an Arabidopsis transcription factor, DREB2A, involved in drought-responsive gene expression, *Plant Cell* ,18, 1292-1309.

[104]  Schramm, F., Larkindale, J., Kiehlmann, E., Ganguli, A., Englich, G., Vierling, E., & von Koskull-Döring P., (2008). A cascade of transcription factor DREB2A and heat stress transcription factor HsfA3 regulates the heat stress response of *Arabidopsis. Plant J.* 53, 264-274.

[105]  Díaz-Martín, C., Almoguera, P., Prieto-Dapena, J., Espinosa, J. M., & Jordano,R.A., (2005) Functional interaction between two transcription

factors involved in the developmental regulation of a small heat stress protein gene promoter, *Plant Physiol.* 139, 1483-1494.

[106] Narusaka, Y., Nakashima, K., Shinwari, Z.K., Sakuma, Y., Furihata, T., Abe, H., Narusaka, M., Shinozaki, K., & Yamaguchi-Shinozaki, K., (2003). Interaction between two cis-acting elements, ABRE and DRE, in ABA-dependent expression of Arabidopsis rd29A gene in response to dehydration and high-salinity stresses. *Plant J.,* 34, 137-148.

# INDEX

## K



## L



## M

## R





## S

## T